AF552835

# PLANT PRODUCT BIOTECHNOLOGY

# PLANT PRODUCT BIOTECHNOLOGY

*By*

**Dr. P.R. Yadav**
*Lecturer*
*Department of Zoology*
*D.A.V. College*
*Muzaffarnagar (U.P.)*

*&*

**Dr. Rajiv Tyagi**
*Department of Zoology*
*M.M. College*
*Modi Nagar (U.P.)*

DISCOVERY PUBLISHING HOUSE
NEW DELHI-110002

*Published by:*
**Namit Wasan**

**DISCOVERY PUBLISHING HOUSE PVT. LTD.**
4383/4B, Ansari Road, Darya Ganj
New Delhi-110 002 (India)
*Phone* : +91-11-23279245; 23253475; 43596065
*E-mail* : discoverybooksindia@gmail.com
discoverypublishinghouse@gmail.com
namitwasan9@gmail.com
*web* : www.discoverypublishinggroup.com

***Edition:*** **2020**

**ISBN: 978-81-8356-083-2**

***Plant Product Biotechnology***



*Printed at:*
Infinity Imaging Systems
Delhi

# Preface

The present title "Plant Product Biotechnology" focus on new biotechnological processes to expand and develop the range of products that originate from plants. As a result of the growing understanding of the process involved, biotechnology is also helping to reduce any adverse impact on the environment. The present title provides a comprehensive review of specialist research directed towards efficient and environmentally sensitive use of plants. The whole gamut of the subject has been described in a simple and systematic manner suited for readers who have not been exposed sufficiently to the subject.

To make the work more comprehensive and informative, the author has consulted many authoritative books, research journals, abstracts, monographs etc. He is grateful to all those great scholars whose work are cited or substantially reproduced.

There can be no claim to originality except in the manner of treatment and much of the information has been obtained from the books and scientific journals available in the different libraries.

The author expresses his thanks to his friends and colleagues whose continue inspirations have initiated him to bring out this book.

The author is painfully aware of the shortcomings, errors and misprints that have crept in, and shall be greateful to receive suggestion for improvement of the next edition from all the readers.

The author expresses his gratitude to Mr. Wasan and staff of M/s Discovery Publishing House for their whole hearted co-operation in the publication of this book.

Author

# Contents

# Chapter 1

# INTRODUCTION

The development of useful *ex situ* germplasm collections, whether for genetic resource conservation or in support of crop improvement programmes, requires that plants be collected from various sites around the world and introduced to new locations. A goal of many germplasm repositories is to represent the global genetic diversity of plant genera in their charge. One of the risks associated with collection of plant germplasm, especially from wild sources, is the inadvertent introduction of diseases or other pests along with plants or seeds. Some of the world's most destructive plant diseases have been the result of the accidental introduction of exotic pathogens during the importation of plant materials. Examples of such importations include downy mildew of corn and sorghum caused by *Peronosclerospora sorghi*, white pine blister rust (*Cronartium ribicola*) chestnut blight (*Cryphonectria parasitica*, and more recently karnal bunt of wheat (*Neovossia indica*), sharka disease of *Prunus* sp. and tomato spotted wilt virus. Virus and virus-like diseases pose a special challenge as they often symptomless in infected plants and require specialized tests to determine their presence. As plant breeders attempt to broaden the genetic base of our agricultural and horticultural crops there are increasing efforts to introduce new genetic material from areas of origin of the species involved. These introductions are often land races or wild species that are not closely examined for the presence of germplasm-borne pathogens.

## SAFE MOVEMENT OF GERMPLASM

Quarantines have become the primary strategy for preventing the international movement of viruses and other pests along with plant material. Finding a balance between excluding pests through the use of quarantine measures and introducing germplasm for the improvement of agricultural production is often a challenge. The destruction of plant material found to be infected with a pathogen or the many years of testing often required before material can be released from quarantine, has frustrated farmers, horticulturists and breeders anxious to incorporate new genetic resources into their programmes.

*In vitro* culture and molecular diagnostic techniques are helping to expedite the detection and elimination of pathogens from plant germplasm. The protection of virus tested clones in certification programmes helps to safeguard valuable plant material, and prevent the reintroduction of pathogens into clean stock. With increased international movement of germplasm the challenge to present introduction of exotic pathogens is also increased. The ease of air travel provides a multitude of opportunities for the introduction of a 'favourite' fruit, vegetable or flower; however, this introduction can lead to serious consequences if an exotic plant pathogen is inadvertently introduced at the same time. It is believed that this is how necrotic strain of potato virus Y ($PVY^N$) was introduced into eastern Canada, resulting in millions of dollars in losses to the seed potato industry in that region.

## Quarantine

Quarantines are the first line of defence against the movement of economically important plant pests between and within countries. Most countries have enacted quarantines to prevent the economic losses that result from the introduction of exotic insects or diseases Introduced pests have been responsible for devastating losses of native plants and cultivated crops. Quarantines are not only used between countries, but are also an important administrative tool for preventing the spread of diseases within countries. In the United States, for example, certain states with pine based timber industries restrict the importation of *Ribes* species, which are an alternate host for *Cronartium ribicola* (white pine blister rusts). The state of Oregon restricts the importation of *Corylus* species (hazelnut) from the eastern two-thirds of North America as well as from several Oregon countries to prevent the further spread of *Anisogramma anomala* Peck. Provinces in western Canada controlled the movement of potatoes from the Maritime Provinces during the early 1990s to protect their seed potato industries from the introduction of $PVY^N$.

## International Cooperation

The International Plant Protection Convention (IPPC) was adopted by a conference of the Food and Agriculture Organization (FAO) of the United Nations in 1952 with the purpose of 'securing common and effective action to prevent the spread and introduction of pests of plants and plant products and to promote measures for their control' (FAO, 1998). As of 1996, 105 countries had agreed to abide by the IPPC. The IPPC directs member countries to develop procedures for issuing phytosanitary certificates, and encourages international cooperation to prevent the spread of pests while minimizing interference with international trade. While most individual countries maintain their own plant quarantine regulations, several regional plant protection organizations have been formed by IPPC member countries in an attempt to provide regional consistency in regulations, and cooperation in the development of plant certification schemes. Contact information for these regional organizations can be found at the FAO website. Three additional regional organizations administer plant protection programmes not associated with the IPPC. The FAO has defined two categories of quarantinable pests. An 'A-1' pest is not yet present in an area and is the most significant from a quarantine standpoint. An 'A-2' pest may be present in an area but is not widely distributed. Quarantine regulations are generally more rigorous for A-1 than for A-2 pests.

### *Europe*

The European Plant Protection Organization (EPPO) makes recommendations to its 39 member countries in Europe and the Mediterranean basin regarding plant quarantine issues related to the movement of commodities.

The 15 nations of the European Union (EU) abide by EU phytosanitary laws, many of which are consistent with EPPO recommendations. EPPO has published a number of certification schemes, for example for strawberries and for fruit trees, with specific suggestions for selection, production and maintenance of nuclear stock, and guidelines for pathogen testing and sanitation.

### *United States*

In the United States, importation of nursery stock, plants, roots, bulbs, seeds and other plant products is controlled by foreign quarantine regulations. The Animal and Plant Health Inspection Service (APHIS), an agency of the US Department of Agriculture, is charged with enforcing plant quarantine regulations. For most plant genera, seeds have fewer restrictions than vegetative materials. Plant material imported for propagation will fall under one of the following categories, depending on the plant species and the country of origin:

*Restricted*

These plant propagules can be imported by any individual in any quantity but are subject to inspections for quarantine pests.

(a) Many plant species are not mentioned in the US quarantine regulations. Such plants or seeds may be inspected for visible evidence of diseases or insects at the port of entry, and if the propagules appear healthy, they are released to the importing individual.

(b) Individuals must obtain a permit to import other plants. These plants or seeds must be inspected at the port of entry and if there is no sign of insects or disease, they are released to the permit holder.

(c) Plant materials which are at risk of harbouring more hazardous pests must be imported with a permit, inspected at an inspection station, and then grown according to certain 'post-entry' conditions. Post-entry restrictions generally require that plants be grown at an approved site, kept a certain distance away from other plants of the same genus, and inspected by an agricultural official several times during two growing seasons.

*Prohibited*

Certain vegetatively propagated plants, including many of the fruit and nut crops, cannot be imported directly by private individuals or organizations in commercial quantities. These plants can only be introduced in small quantities through a government approved quarantine facility, where they must be tested for economically important pathogens (especially viruses) before being released for general propagation and dissemination.

## International Guidelines

The International Plant Genetic Resources Institute (IPGRI) is another organization associated with FAO which promotes the conservation and use of genetic resources. In recognition of the phytosanitary hazards involved in the international movement of plant germplasm, IPGRI has funded a number of research programmes and conferences to develop appropriate technologies for reducing the risks of moving pathogens with plants. A series of crop-specific handbooks have been published by IPGRI to provide technical guidelines for the safe movement of a number of economically important crops plants.

These guides have especially targeted vegetatively propagated crops which carry a higher risk of carrying virus diseases and they suggest appropriate methods for excluding these diseases during germplasm exchange. Some general recommendations to promote the safe movement of germplasm include:

1. Germplasm should be obtained from a safe (pathogen tested) source.
2. Movement of seed or pollen poses less risk than does movement of plants.
3. *In vitro* cultures pose less risk than does vegetative plant materials potted in soil.
4. *In vitro* material should be derived from pathogen tested sources or tested for viruses known to occur in the country of origin.
5. Transfer of germplasm should be planned in consultation with quarantine authorities and a relevant indexing laboratory.
6. Vegetative material should be subjected to full quarantine measures.

Reviews of successful certification programmes have recently been published for potatoes, grapes, ornamental plants, deciduous fruit trees, citrus, and strawberries. Common features of certified stock schemes involve:

1. Listing the pathogens of concern.
2. Identifying appropriate methods for detecting the pathogens.
3. Implementing strategies for eliminating pathogens from infected plants.
4. Protecting foundation or nuclear stock from reinfection.
5. Verifying that plants are correctly labelled or are 'true-to-type'.
6. Preventing reinfection during subsequent propagation and distribution to commercial producers.

IPGRI recommends the development and use of 'broad-spectrum' virus detection techniques for quarantine indexing. Such tests may not identify specific pathogens, but rather will detect members of a larger group such as all plant phytoplasmas, or all members of a virus family.

## VIRUS DETECTION

Accurate disease diagnosis combined with sensitive, rapid and early detection of plant viruses is critical for effective management of most crop systems. However, in plant germplasm collections detection of a wide range of viruses, some of which may not be known, is required generally for a small number of samples. There have been several major improvements in virus detection over the past two decades. Serological detection was greatly improved with the introduction of ELISA to plant virus detection in the mid-1970s. This was enhanced further with the development of monoclonal antibody technology and its application to a large number of plant viruses in the 1980s. Similarly, nucleic acid hybridization has been used successfully for detecting many plant viruses and is the preferred method for detection of viroids. Cloning of plant viral nucleic acids and the development of non-radioactive detection methods have increased the polymerase chain reaction (PCR) has greatly improved the sensitivity and utility of hybridization and other nucleic acid based assays. Immunocapture PCR combines the advantages of serology and PCR into a very sensitive method of detection.

In germplasm repositories it is often necessary to test for viruses in accessions that are from diverse geographical locations and that may be infected with unknown viruses. Tests for common viruses known to occur in a genus should be carried out. However, repositories have the added need to determine whether these materials may be infected with yet undescribed viruses. This necessitates the use of broad spectrum techniques such as mechanical transmissions to herbaceous indicators, electron microscopy, double-stranded RNA analysis, grafting onto indicator hosts or looking for viral inclusions.

Monoclonal antibodies, nucleic acid hybridization and PCR provide the potential for the development of diagnostic reagents with desired specificities. Diagnostic reagents that detect all or most members of a virus group would be very useful in germplasm repositories. There are monoclonal antibodies that recognize most members of the potyviridae or multiple luteoviruses and oligonucleotides that can be used in PCR to amplify sequences from most luteoviruses or geminiviruses.

**Requirements for Detection and Diagnosis**

The requirements for specificity and sensitivity of detection will vary in different situations. Germplasm repositories and clean plant programmes want to ensure that their 'nuclear' material is free of known viruses. In this situation, where much depends on the virus status of relatively few individual plants, several tests should be employed and the virus status of every plant may determined. While serological or nucleic acid-based tests often can be employed for detection of well characterized viruses known to occur in the crop, tests that detect a wide range of viruses are especially desirable. Mechanical transmission to selected herbaceous indicator plants, electron microscopic examination of leaf dips, double-stranded RNA (dsRNA) analysis and/or grafting may be desirable to ensure that the germplasm or 'nuclear' material is virus-free. 'Nuclear' material refers to the few plants that are the basis of all plant material in a clean plant propagation scheme and may also be referred to as mother block, foundation, or elite material. New material being added to virus-free collections may be put through a virus eradication programme prior to testing. This enhances the likelihood that the material will be free of viruses not reported previously in the crop and free of new strains of a virus which may not be detected with available antibodies or nucleic acid probes.

Germplasm repositories often have permits to bring in material to meet their mandate without having the plants go through a plant quarantine station. Therefore, their virus detection programmes should be similar to those of plant quarantine. In plant quarantine it is necessary to ensure that plant material is free of restricted viruses prior to its release to industry or breeding programmes. Quarantine programmes often work with only a few plants of a given genotype but the level of indexing on these few plants is as complete as possible. Maximum sensitivity is the goal of virus detection in quarantine programmes. These programmes often use an agreed upon test for each specific virus. The test may be grafting or mechanical transmission onto specific indicator plants, an ELISA, hybridization assay or PCR for specific viruses. Detection of viruses of woody plants will often require graft transmission tests to indicator plants. When grafting is required for a single virus in a crop it might be more efficient to employ graft transmission tests for all viruses of quarantine significance since the work is already being done for the one virus. Thus, quarantine facilities may not opt for the newer laboratory techniques until they are available for all quarantinable viruses capable of infecting that crop.

Quarantine officials may also require extreme specificity in cases where a severe or resistance breaking strain of a virus occurs in some countries but not others. When there are several sources of plant material it might be easier to destroy all material that is infected with a particular virus rather than trying to determine which strain of a virus is present. There is a resistance breaking strain of raspberry bushy dwarf virus (RBDV) that occurs in Europe and is not known to occur in North America. The only way to differentiate the two strains is to graft onto resistant varieties of raspberry. Both strains are readily detected, but cannot be differentiated serologically and in practice any material being imported to North America from Europe that tests positive for RBDV is destroyed rather than risk introducing the resistance breaking strain of the virus. Recently, reverse transcription combined with PCR (RT-PCR) has been developed to differentiate these two strains of RBDV and as this test is improved it may be possible to reliably detect resistance breaking strains of RBDV.

## Serological Detection

Enzyme-linked immunosorbent assay (ELISA) and dot immunobinding assay (DIBA) are currently the most widely used methods of serological detection for plant viruses. There have been relatively few changes in ELISA or DIBA for virus detection since the application of monoclonal antibodies (McAbs). ELISA has been reviewed elsewhere and the general outline will not be covered here. Rather, some of the principles of the assays will be discussed. These assays are carried out on a solid phase (usually plastic multi-well plates, nitrocellulose membranes or filter paper where each component of the test is applied successively and the reaction between virus and antibody is detected by enzymatic hydrolysis of a substrate that results in a colour change or light emission.

Assays carried out on nitrocellulose or filter paper are referred to as DIBA. The DIBA is about as sensitive as ELISA and offers the advantage that sample preparation can be very simple. A few microlitres of sap can be spotted onto the membrane or a freshly cut edge of a leaf, petiole or stem can be pressed gently against the absorbent membrane. The latter approach is referred to as tissue blotting and also provides some information on the location of the virus in the tissue, e.g. phloem. The other advantage of DIBA is that it is readily adapted to field situations and applications in areas with minimal laboratory facilities. The membranes can be taken to the field and plant tissue or insects blotted directly onto the membrane. Both of these assays can be performed without any specialized equipment.

The ELISA procedure is quite adaptable and many variations on the original method have been described. A standardized test protocol should be used in quarantine and certification schemes to ensure consistent results between laboratories or from year-to-year in the same facility. Standardization of an ELISA protocol may include factors such as the manufacturer of a microtitre plate used in the assay, identification of a specific monoclonal or polyclonal antiserum, part of plant to be sampled and at what stage of plant development sampling should take place, list of buffers to be used at each step of the procedure, the length and temperature of each incubation, how long the substrate should develop before absorbance values are taken, and what should be used as positive and negative controls.

ELISA in plant virology usually has two or three steps that use an antibody. The antibody applied directly to the microtitre plate is usually referred to as the trapping or coating antibody, since its purpose is to selectively bind the antigen of interest to the plate. The coating antibody is applied as purified immunoglobulin G (IgG) diluted in an appropriate buffer (usually carbonate or phosphate buffered saline). The second antibody (often the same source of IgG as the coating antibody) is conjugated with an enzyme in the standard double antibody sandwich (DAS) ELISA and is referred to as the conjugate or detecting antibody. In triple antibody sandwich (TAS) ELISA, the coating is done in the same manner as in DAS-ELISA but the second antibody (primary or detecting antibody) is specific for the antigen of interest and produced in a different animal than the trapping or coating antibody. The primary antibody is then followed by a conjugated antibody (secondary antibody) that is specific for antibodies produced by the animal that was the source of the primary antibody. For example, if the primary antibody was a McAb produced in a mouse, then the secondary antibody might be a rabbit-anti-mouse conjugate. We prefer a conjugate that is made in the same host species as the coating antibody to prevent cross reaction between conjugate and coating antibody.

When using monoclonal antibodies (McAbs) for virus detection or diagnosis, the substrate can usually be incubated overnight to increase the sensitivity of the assay. In this way, McAbs can be used to detect virus in individual aphids using a standard ELISA protocol. In our laboratory, we routinely take absorbance readings 1-2 hours after adding substrate and again after overnight incubation at room temperature. The most important aspect of developing an assay is to maximize the absorbance of infected/ healthy samples. A good polyclonal or monoclonal antiserum can be used to develop an assay that does not require statistical analysis to differentiate between known positive and negative samples. If such a test is available, samples that give borderline results should be retested. It must be remembered however, that even with the best of detection methods, recently infected samples may well give a borderline or negative result. The application of statistics to determine when an absorbance value in ELISA represents a positive result has been present elsewhere.

## Nucleic Acid Based Assays

Detection of pathogens by nucleic acid hybridization is based on the specific pairing between the target nucleic acid sequence (denatured DNA or RNA) and a complementary nucleic acid probe to form double-stranded nucleic acids. Thus, either RNA or DNA sequences may be used as probes. For detection of plant viruses, hybridizations are usually carried out on solid filter supports where the target nucleic acids are immobilized and the labelled nucleic acid probe is allowed to hybridize to them. Nitrocellulose and charged nylon are the most commonly used filters for hybridization. Nylon based filters are easier to handle and can be rehybridized several times. The use of nucleic acid hybridization assays for virus detection has been reviewed recently. The greatest improvements in hybridization assays in recent years have been the advances in non-radioactive detection systems. Several methods of labelling nucleic acids are available. Incorporation of modified nucleotides such as biotin-11-UTP, or digoxigenin tagged UTP, can then be detected by streptavidin or an anti-digoxigenin antibody, respectively. Digoxigenin-labelled dUTP appears to be the most widely used non-radioactive tag for labelling probes for detection of plant viruses. Digoxigenin is linked via a spacer arm to dUTP and then incorporated into probe with the same enzymes used to make $^{32}$P-labelled probes. Antibodies specific for the digoxigenin and conjugated to an enzyme (usually alkaline phosphatase or horseradish peroxidase) are then used to complex with the bound probe. Either a precipitating or a light emitting substrate is then used to visualize the presence of the probe. Many biotechnology supply companies now marked kits for tagging probes with non-radioactive labels.

The sensitivity of detection of plant viruses by nucleic acid hybridization is roughly similar to that of ELISA. Hybridization assays were more sensitive than ELISA with subterranean clover stunt virus (SCSV) when purified virus was used; however, when infected tissues were used, ELISA was found to be more sensitive: the pathogen was detectable at a sap dilution of 1/625 while hybridization assays had a dilution limit of 1/125. The sensitivity of nucleic acid hybridization and ELISA were similar with barley yellow dwarf virus (BYDV) and cherry leaf roll viruses.

Another consideration in the choice of detection procedure, is the ease of carrying out an assay. If two methods of detection are sensitive enough to meet the needs of an experiment, then the method that is simpler to carry out will probably be used. The expertise of the worker will determine which test is simpler. Someone who has extensive experience working with nucleic acids will be more inclined to use PCR or hybridization assays rather than ELISA; the converse is true for someone who is more familiar with serology than nucleic acid based tests.

There are many viruses of woody plants that have only been described in terms of symptomatology. It is likely that nucleic acid based detection will be available before serological methods are developed for these viruses. Developments in cloning viral specific dsRNAs of plant viruses make it possible to readily make cDNA from dsRNA templates. Thus, for viruses where dsRNA can be extracted from diseased tissues, probes for nucleic acid hybridization or sequence information required to develop a PCR test are realistic short-term goals. Once the sequences are known, it will be possible to prepare antibodies to the coat proteins of these viruses for use in serological assays.

In the case of phytoplasmas, several groups have described universal oligonucleotides for amplification of phytoplasma specific

DNA. Since this group of pathogens cannot be cultured, are not easily transmitted by grafting or mechanically, and can be difficult to detect by microscopy, the PCR based test is the preferred method for detecting phytoplasmas.

## Detection Based on more Traditional Methods

Serological or nucleic acid based diagnostic tools are not available to detect many of the viruses of woody plants because it has not been possible to purify the viruses. It is still necessary to carry out graft transmissions or mechanical inoculations onto herbaceous hosts for these viruses. Attempts to improve quarantine and certification schemes for tree fruit or small fruit crops is limited because these tests are labour intensive, and especially with grafting, require substantial amounts of space to grow test plants. It may be several years after grafting before the results of the test are obtained.

The introduction of new germplasm from collections made in the wild presents the possibility that undescribed viruses will be encountered. New viruses encountered will not be on any quarantine list but may still pose a biological risk. This material should be assayed with a broad spectrum test such as mechanical transmission, grafting or dsRNA analysis. Quite often we tend to be concerned about the viruses with which we are familiar and ignore those that have not yet been described.

## Significance of a Test Result

In situations where a test result has significant biological or economic impact, one should use more than one type of test to confirm the presence of absence of a virus. A recent incidence of $PVY^N$ in Canada and its implications for shipping seed potatoes to the USA is an excellent example of where test results had considerable economic impact and a confirmatory test could potentially have saved a lot of money and possibly prevented lawsuits. In this case the effect of mixed virus infections confounded the bioassay results on tobacco leading to a large percentage of false positives. As the movement of agricultural products between countries increases, quarantine restrictions based on plant pathogens will continue to be important. For example, several *Prunus* spp. were found to be infected with virus isolates which cross-reacted with plum pox potyvirus (PPV) antisera in ELISA. In this instance, the impact of the test result was very significant since plum pox is an A-1 quarantine status virus not known to occur in North

America. Plum pox is a member of the potyviridae that produce diagnostic inclusions in infected hosts. The prunus virus isolates did not induce these inclusions; their coat proteins were a typical of members of the potyviridae and RT-PCR tests with oligonucleotides specific to the 3' or 5' non-coding regions of plum pox virus did not amplify any fragments. It is now thought that these prunus virus isolates are not plum pox virus, nor are they members of the potyviridae despite their reaction with antisera specific to plum pox virus. Reliance solely on the initial test results would have had very serious implications.

Another important consideration when using diagnostic tests that are not based on biological activity is the significance of the result. For example, many people are aware that PCR has been used to amplify DNA from insects embedded in amber for millions of years. This positive 'test' for insect DNA does not show that the insect is living. A similar situation could arrive when using laboratory tests to index plants for viruses. It has been shown that biologically active prunus necrotic ringspot virus is slowly lost from *Prunus pennsylvanica* seed. Would ELISA, PCR or hybridization give positive results when biologically the virus was no longer seed transmitted? In the case of biological tests, it is also important to know that the symptoms observed are due to the virus in question rather than a mixed infection as was the case with some of the $PVY^N$ testing mentioned above.

The legal implications of a decision to refuse a shipment of plant product based on a positive test that does not consider biological activity is an issue that needs to be considered. If a shipment of grain that has been fumigated with methyl bromide is assayed for a specific fungal pathogen by PCR, it would be possible to get a positive result based on non-living fungal tissue. Similarly, the results of a positive ELISA test for plum pox potyvirus could be used to turn back a shipment of *Prunus* planting stock, when in fast the material was free of plum pox potyvirus. We must remember to consider the biology of the pathogen and host rather than rely on a band in a gel or yellow colour development in a well of a microtitre plate.

## Production of Pathogen-free Plants

Most plants that are vegetatively propagated from a virus-infected plant will be infected with the same viruses. These clonal plant populations can become infected with additional viruses if exposed to virus-carrying vectors such as nematodes or aphids. Many heirloom fruit cultivars have been clonally propagated for centuries and are universally virus infected, although the viruses may be latent or symptomless during much of the growing season. Pathogen-free plants can sometimes be identified by careful indexing, but often the only way to obtain healthy foundation stock for important clonally propagated plant cultivars is to subject them to virus-elimination therapy.

Exposure of growing plants or plant parts to elevated temperatures (heat therapy or thermotherapy), to apical meristem culture, or to antiviral chemicals (chemotherapy) has been used to eliminate viruses from infected plants. Improved virus elimination can often be achieved by combining these various techniques. The treated plants generally are not cured during therapy, but rather new plants are propagated from shoot tips or apical meristems following treatment. Plant tissue culture offers a convenient system for exposing plants to controlled temperatures or to controlled concentrations of antiviral chemicals. *In vitro* therapy also provides already-sterile plant tissue from which meristems can be dissected with no need for additional surface sterilization.

**Heat Therapy**

While the exact effect of heat therapy on plant viruses is not well understood, it is known that replication of many viruses is significantly reduced at elevated temperatures. The production of virus-encoded movement proteins and coat proteins may also be temperature sensitive. These proteins are involved in cell-to-cell movement of viruses through plasmodesmata and long distance movement through the plant vascular system. Disruption in the production or activity of these proteins may also play a role in the effectiveness of heat therapy.

Nyland and Goheen (1969) reviewed the early history of using heat to eliminate viruses and other pathogens from infected plant material. Brief exposure of plants or plant parts to hot water at temperatures ranging from 30° to 70°C has been used successfully to eliminate many pathogens from infected plants, but nearly all of the pathogens eliminated by hot water baths were later discovered to be phytoplasmas and not viruses. Hot water may be a useful technique for sanitizing plant parts to eliminate arthropods, fungi, bacteria and phytoplasmas, but it has not been particularly useful for eliminating viruses.

Most modern virus therapy programmes involve growing whole plants or *in vitro* cultures at temperatures close to the threshold of normal plant growth. For most plants this is between 38° and 40°C. Mink et al. (1998) reviewed the effect of elevated temperatures on viruses and on plant physiology. Reduced synthesis of RNA at 40°C has been shown for several viruses, including tobacco mosaic virus (TMV) and cowpea chlorotic mottle virus (CCMV). At elevated temperatures synthesis of ssRNA stopped immediately for both of these viruses and synthesis of dsRNA stopped more gradually. When plants were returned to normal temperatures of 25°C, resumption of viral RNA synthesis lagged behind plant RNA synthesis by 4-8 hours for CCMV and by 16-20 hours for TMV.

Infected plants may be subjected to constant therapy temperatures, however alternating temperatures are commonly used to improve host survival. The lag in resumption of viral RNA synthesis behind that of the host plant may explain why alternating-temperature heat therapy is perhaps more successful than constant-temperature therapy. Plant survival is improved during alternating-temperature therapy. We have successfully eliminated many viruses from numerous fruit, nut, and oil crop genera (*Corylus*, *Fragaria*, *Humulus*, *Mentha*, *Mespilus*, *Pyrus*, *Rubus*, *Ribes*, *Sorbus*, *Vaccinium*) by growing either potted plants or *in vitro* plantlets at temperatures that alternate every four hours between 30° and 38°C. After two to four weeks of heat therapy, apical meristems are dissected and established *in vitro*. Once these meristem derived plants are rooted and established in soil, they are allowed to go through a natural dormant period and retested for viruses during the following growing season. The dormant period allows viruses, that may have been reduced to undetectable levels by therapy, to build up in the plant prior to indexing.

While typical virus therapy involves growing an infected plant at elevated temperatures prior to meristem culture, cucumber mosaic and alfalfa mosaic viruses have been eliminated by subjecting dissected meristems to heat therapy after they were removed from infected plants. Cold therapy rather than heat therapy, followed by apical meristem culture, has successfully eliminated several viruses from infected plants and cold therapy has been particularly effective in the elimination of viroids, some of which are quite resistant to elevated temperatures. Plants or *in vitro* cultures are typically grown at temperatures between 4° and 7°C for one to six months prior to removal of meristems to eliminate viroids including potato spindle tuber, chrysanthemum stunt, and apple scar skin viroids.

While there seems to be little agreement on the importance of humidity or light levels during therapy, it has been shown that elevated $CO_2$ enhances survival of some plant species at elevated temperatures. Blueberry (*Vaccinium* spp.) plants, which normally are difficult to keep alive at 38°C were able to survive for extended periods at 40°C when the $CO_2$ level was increased to 1200 ppm, 3-4 times the normal concentration. We have observed improved survival of the infected host during *in vitro* heat therapy if the plantlet is cultured in a heat-sealed gas-permeable plastic pouch rather than a test tube or other culture container. These plastic containers allow some gas exchange but are impervious to the exchange of moisture. Improved plant condition may be due to a decrease in fungal or bacterial contaminants, better water retention in the medium, or elevated $CO_2$ levels in these containers.

Viruses vary in their susceptibility to heat therapy. Apple mosaic ilarvirus, blueberry scorch carlavirus, and several of the mosaic viruses of raspberry are readily eliminated following therapy times of only 10-20 days and in the case of apple mosaic, following propagation of relatively large shoot tips over a centimetre in length. Other viruses such as raspberry bushy idaeovirus, tobacco streak ilarvirus and apple stem grooving capillovirus persist in some meristems smaller than 0.5 mm dissected after 3-4 weeks of heat therapy.

## Meristem Tip Culture

Virus elimination has been documented for a number of virus-host combinations following prolonged *in vitro* culture of the infected host plant, and in many additional instances following excision and culture of apical meristems. Faccioli and Marani (1998) recently reviewed the various roles that *in vitro* culture has played in elimination of viruses from infected plant material. The mechanisms for virus elimination during routine tissue culture are uncertain, however the presence of plant growth hormones, the physiological response of explants to repeated injuries, the induction of inhibitors or the disruption of enzymes needed for virus replication are several possibilities. Cherry leaf roll virus and arabis mosaic virus for example, have disappeared from infected plants after several months of growth *in vitro*, yet tobacco ringspot, cucumber mosaic and potato virus X are persistent.

It has been suggested that many viruses are unable to infect the apical meristem of a growing plant and that a virus-free plant can be produced if a small enough piece of apical tissue is propagated. Facciolo and Marani (1998), however, cite a number of studies where electron microscopy and fluorescence-linked antibodies have documented the presence of more than a dozen different viruses in apical dome tissue of assorted plant species. Yet plants regenerated from these infected meristems are often free of the viruses. The larger the size of an excised meristem the better the chance that it will survive *in vitro* culture, and the smaller the meristem size, the more likely it will be virus-free. The goal of many virus elimination programmes is to grow a meristem consisting of the apical dome and a pair of leaf primordia. Depending on the plant genotype, this shoot tip should be between 0.2 and 0.8 mm in length. Distribution of a virus within a plant may be uneven, especially towards the shoot tips, and a virus-free plant may be produced by random propagation of enough buds or shoot tips. Experience with a particular crop and the viruses that infect it will determine the size of the meristems that can be established *in vitro*, and how many meristems must be grown to have a reasonable probability that one of them will be virus-free. A single, virus-free, true-to-type plant is all that is necessary to produce a population of healthy plants.

Some woody plants are difficult to establish in *in vitro* culture from meristems, or difficult to root. These difficulties have been overcome in *Citrus*, *Malus*, and *Prunus* by 'micrografting' shoot tips or apical meristems onto *in vitro* grown seedling rootstocks, which are later transplanted to soil after the grafts have become established. Pears which grew easily from *in vitro* meristems, but which were difficult to root, could be removed from tissue culture when they had elongated to about 1 cm and successfully cleft-grafted onto potted seedlings. Selection of rootstock seedlings with distinct leaf colour or morphology aids in differentiating micrografts from rootstock sprouts as they grow out.

Although it is possible to eliminate viruses from plants following meristem tip culture alone, this procedure is almost always combined with heat therapy or chemotherapy to increase the likelihood of success. The combination of heat therapy followed by apical meristem culture has become the foundation for many virus elimination programmes at germplasm repositories, research institutes and commercial plant nurseries around the world.

**Chemotherapy**

Unlike fungi and bacteria, viruses cannot be eliminated from infected plants by protective chemical sprays. Several chemicals, however, can be used in the same manner as heat to inhibit the replication or movement of viruses thus producing a region of virus-free tissue. The chemicals can either be sprayed on growing plants or incorporated into tissue culture media. Following a period of chemotherapy, shoot-tips or apical meristems are excised and propagated as in heat therapy. While it may be possible in some instances to cure an infected plant, the cost of these chemicals precludes the application of chemotherapy to field trees.

Many of the antiviral chemicals that have been used for plant virus chemotherapy are synthetic nucleotide analogues that had previously shown some effectiveness against animal viruses. The concentrations that are required during chemotherapy to inhibit virus multiplication are very close to the concentrations which are toxic to the host plant. The guanosine analogue ribavirin (Virazole 1-D-ribofuranosyl-1,2,4-triazole-3-carboxamide), and the uracil analogue DHT (5-dihydroazauracil) are two substances which are particularly effective at inhibiting many different plant viruses. Hansen (1988) reviewed applications of chemotherapy to virus infected plants. In early studies, antiviral substances were injected into stems or applied to whole plants as foliar sprays or root drenches. More recent work involved the incorporation of antiviral materials into tissue culture media leading to the elimination of viruses from many plants including peanut (*Arachis*), orchid (*Cymbidium*), potato (*Solanum*), and strawberry (*Fragaria*). Tissue culture methods permit more control over the concentration of chemical agents, the period of exposure, and the ability to combine chemotherapy with heat therapy or apical meristem culture. Combinations of more than one chemical are also more easily evaluated using tissue cultures.

The modes of action of the various plant virus chemotherapies are not well understood. Ribavirin, perhaps the best studied, has been shown in animal viruses to interfere with capping at the 5' end of viral mRNA and many of the ribavirin-sensitive plant viruses also have a 5' capped mRNA. While chemotherapy is effective for eliminating a diverse array of viruses from many different plant species, the chemicals used, the effective concentrations, and the phytotoxicity levels vary greatly depending on the plant genotype. The possibility of genetic mutations when plants are exposed to antiviral chemicals poses risks that have not been adequately studied. Heat therapy poses a much lower risk of genetic change to treated plants, and the procedure is consistent regardless of the virus or the plant host species. Chemotherapy may supplement the other methods of plant virus elimination, but is not likely to replace them, except in situations where viruses are not easily eliminated by heat and meristem culture.

Not all plants resulting from therapy will be pathogen-free. Each plant must be subjected to follow-up indexing to confirm the virus status, and plants should be grown out and examined for trueness to type before being offered for commercial production. There are abundant opportunities for plants to become mislabelled during the various stages of propagation, heat therapy, *in vitro* culture, and re-establishment. Concerns have also been raised about the possibility of genetic mutations or selection of somaclonal variants especially following *in vitro* culture procedures which involve certain plant growth hormones. Such changes are less likely if regeneration from undifferentiated tissues or callus is avoided during *in vitro* culture. Heat therapy and apical meristem culture not only can eliminate viruses that have been previously detected, but may also eliminate exotic or latent viruses whose presence is unknown, as well as contaminating organisms such as bacteria, fungi, or phytoplasmas.

## CONCLUSIONS

Germplasm collection present a unique set of problems with regard to prevention of movement of plant viruses. Due to the nature of the plant material collected, there is always the possibility of introducing completely unknown viruses. A more complete set of virus tests needs be carried out on these plants compared to other virus testing programmes such as certification programmes dealing with a known set of viruses. Since germplasm collections deal with a broad range of genetic materials, sensitivity to thermotherapy or chemotherapy, and the ability to culture the plants *in vitro* may be more variable than in certification programmes where the genetics of a crop is more uniform.

Once pathogen-free germplasm is identified, whether through indexing of collected accessions, or through a combination of therapy and indexing, the plant material should be protected from reinfection. In conventional plant gene banks and certification programmes this involves isolating growing plants from virus vectors. Potted plants maintained in insect-proof enclosures will be protected from spread of viruses by insects and nematodes. Removal of flowers and exclusion of pollinators such as honeybees will prevent the movement of pollen-borne viruses. Conservation of clonal plant germplasm *in vitro* should assure that plants are protected from becoming infected with pathogens, including viruses, provided the source of the *in vitro* plants is virus-free.

# Chapter 2

# Molecular Tools

The small-grained cereals considered in this chapter, namely wheat, barley and oats, are all natural inbreeders, i.e. any one plant is more likely to reproduce with itself rather than with a neighbour. This means that, once a potential plant variety is homozygous, there is no residual variation and it will reproduce itself exactly from one generation to another. Breeding programmes for the small-grained cereals generally follow a pedigree selection scheme with minor variations in detail. At the beginning, breeders cross parents which complement each other for desirable characteristics to produce the F1 generation which will be genetically uniform. The F1 generation naturally self-pollinates to eventually give rise to a population of inbred lines that will contain $2^n$ different genotypes, where *n* is the number of potential different inbred lines is $3^{20}$ or 1,048,576–a large number that would occupy about a quarter of a hectare at UK commercial sowing rates if each line was represented by just one seed. To be sure of generating one particular gene combination, the population size would need to be larger still and when one considers that a breeder will make several hundred crosses a year it is abundantly clear that plant breeding programmes cannot accommodate all the possible combinations.

Most breeding programmes are therefore constrained to several hundred thousand F2 plants spread over a range of crosses that a breeder will have judged from parental performance to be the best. Breeders will select the best F2 plants on the basis of easily recognized characters such as disease resistance, height and maturity. Further selection on such characters is practised upon the F3 generation, after which some breeders will then carry out a preliminary yield trial at the F4 generation and suitability for processing such as baking wheat and mating barley will also be assessed. Yield and processing characters are affected to varying degrees by the environment in which the crops are grown so that the ranking of individuals may vary between environments. This is known as genotype × environment interaction and, to get a more complete picture of the potential of breeding lines, further yield trials are carried out a number of sites in one or more subsequent generations. At the end of this series of trials, breeders will have assembled a comprehensive body of data upon their remaining selections and will enter the best into official trials. In the UK, approximately 120 winter wheat, winter barley and spring barley lines are entered into National List trials each year in total. National List trials take two years and are conducted according to set protocols in order to select the best three to ten from each crop to enter Recommended List Trials, which are carried out a number of sites all over the UK.

After a year, sufficient data will have been gathered upon the entries to enable a provisional recommendation to be given to between one and five new varieties from each crop each year, although not even one may be widely grown for more than one or two years. From the above brief outline of a conventional cereal breeding scheme, it can be deduced that such schemes are lengthy, occupying up to ten years to provisional recommendation for the small-grained cereals in the UK and twelve years to make a commercial impact. Bingham and Lupton provide a comprehensive review of a practical winter wheat breeding programme, which is typical of many other cereal breeding programmes. Given the huge investment required to run such a scheme over a 12-year cycle, and the competition provided by others, breeders are keen to exploit alternatives which will either shorten the time-scale or increase the efficiency of identification of elite lines and especially any which combine both elements.

The developments in molecular biological tools over the past 15 years have led to the possibility of direct selection for the genetic constitution, or genotype, of individuals. This is an attractive alternative as it reduces the amount of selection upon character measurement, or phenotype, when the environment and its interaction with genotype can modify the expression of genes.

Conventional plant breeding schemes have been outstandingly successful, contributing an average 1% yearly increase in the yield of barley and wheat. The deployment of molecular biological tools in cereal breeding must either lead to more efficient ways of achieving this rate of progress or lead to increased breeding progress.

**Phenotype**

The phenotype of a line is the result of the interaction of its genes with the environment. Thus a character with a high degree of genetic control is less affected by varying environmental conditions than one with a low degree of genetic control. Phenotypic selection for performance is reliable in the former case but is likely to be unreliable in the latter. Breeders can therefore make selections for the former group of characters in unreplicated trials or nurseries in the early stages of a programme and reduce their populations to more manageable levels for field trials. Examples of characters that breeders select for an early generations are disease resistance, maturity and short straw. Many of these characters are controlled by a single gene of large effect (major gene) and are therefore qualitative characters where phenotypic expression is a good indicator of a line's genetic constitution.

Most important characters such as yield and quality are, however, controlled by a number of genes, each often of small effect. In such cases, because variation is generally continuous and it is impossible to discriminate between the different genotypes the characters are termed quantitative. Each individual gene often has a small effect, is subject to considerable environmental modification, and genotype × environment interactions may occur. Accurate prediction of such characters from a single unreplicated trial is extremely unreliable and therefore breeders carry out multi-site trials over two or more seasons to get an estimate of a line's phenotypic potential for characters such as yield and quality.

**Genotype**

The genotype defines the exact genetic constitution of a line but this is unknown for most characters, as the actions of minor genes do not individually produce discernible effects. Indeed, the action of many major genes, such as major gene disease resistance cannot be recognized without extensive testing with race-specific isolates. This means that breeders are usually completely reliant upon phenotype as a predictor of a line's performance. Knowledge of a line's genotype at individual loci, however, can be used to build up a picture of the relationship between loci. Loci that tend to segregate together are said to be linked and the closer the loci are, the greater the degree of linkage and the fewer the recombinants. Knowledge of such relationships is of value to plant breeders as they can then predict which parental combinations will be most likely to give desirable recombinants.

Furthermore if a breeder wished to find recombinants between particular loci, knowledge of linkage relationships can be used to formulate population sizes necessary to produce as many of such lines as desired. Assembling segregation data on a number of loci enables construction of genetic linkage maps, which can be used in conjunction with phenotypic data to reveal regions of the genome that control a character. These regions are termed quantitative trait loci (QTL) and can also be placed on genetic maps.

## MARKERS

### Morphological, Isozyme and Protein

Major gene variation has been known for a long time with many morphological mutants being described. These were used to construct early genetic maps for barley and permitted the addition of major disease resistance genes as they became known. However, many morphological mutants are deleterious and were of little benefit in identifying regions of the genome controlling economically important characters. The effect of some major genes upon other traits could be analyzed in populations segregating for the gene. Such studies were used to identify deleterious effects upon yield and quality characters at two dwarfing loci used in spring barley.

The deployment of *Hordeum laevigatum* as a source of disease resistance has also led to deleterious associations, probably through introgression of undesirable alleles together with the disease resistance genes. However, such findings became known after the genes had been widely deployed and served to explain observed results rather than advocate the selection of alternative genes. Isozyme markers have also been proposed to augment selection of alternative genes. Isozyme markers have also been proposed to augment selection and a number of loci have been included on classical genetic maps of barley and wheat. Some isozymes have been used to select for specific genes, e.g. the endopeptidase gene on wheat chromosome 7D to select for the VPM1 eyespot resistance gene. Analysis of the high molecular weight glutenin sub-units was found to be a good predictor of bread-making quality of wheat and such analyses are now carried out routinely in wheat breeding programmes prior to baking tests, thus increasing the efficiency of selection. In general, the variation at isozyme loci is insufficient to be of great practical use.

## Molecular Markers

Maps constructed from morphological, isozyme and disease resistance genes were the composite picture developed from the integration of a large number of small-scale experiments involving three or four linked loci. The development of methods to assay DNA sequence differences, or polymorphism, permitted the construction of a map of the whole genome from one single population. This has a number of advantages as a much better estimate of the overall order of loci along the chromosomes can be obtained.

### *Restriction fragment length polymorphism (RFLP)*

RFLPs rely on the use of restriction enzymes to digest genomic DNA where polymorphism can arise through mutations to create or remove restriction sites or by sequence deletions or insertions between sites. After digestion with restriction enzymes, the DNA fragments are separated by electrophoresis and labelled DNA probes that recognize specific sequences are then hybridized to the fragments and polymorphisms are detected as length differences. Because different RFLP alleles are recognized on the same gel, the marker system is said to be co-dominant as heterozygotes (or mixtures) produced both allelic bands. The initial use of RFLPs in the small-grained cereals was hampered, to some extent, by the relatively low level of polymorphism that they detected in cultivated germplasm, particularly in wheat. Because RFLP analysis is based upon hybridization assays, comparatively large amounts of DNA of a reasonably high standard of purity are required, which is another limiting factor.

A major advantage of RFLPs, however, is the fact that they provide robust anchor loci that are easily distinguished not only across different crosses within a species but also can be used as inter-specific anchor loci. This latter feature has been of great value in comparative mapping within the *Triticeae*, leading to the ordering of homologous loci across a number of species.

### *Random amplified polymorphic DNA (RAPD)*

The creation of mapping data using RFLPs was laborious, as it required hybridization. The development of molecular marker techniques based on the polymerase chain reaction (PCR) to amplify target DNA segments prior to detection resulted in a quantum leap in the generation of data points. RAPD markers were generated by using random primers in the range of 10-20 nucleotides to detect complementary sites across relatively short distances within the genome. The presence of complementary sits in a genotype resulted in the primer amplitying a DNA fragment under set PCR conditions.

A number of different fragments could be amplified with each primer and separated electrophoretically. The absence of a complementary site results in the absence of a DNA fragment to reveal a polymorphism. This type of marker system is dominant, as a band is either produced or not produced, and the inability to distinguish heterozygotes from a homozygous class is a major disadvantage of the system. Whilst RAPDs could generate data much more quickly than RFLP, they suffered additional disadvantages through difficulties in repeatability, while bands from one cross could not readily be transferred to other crosses. The development of new marker types soon rendered RAPDs obsolete.

### *Amplified fragment length polymorphism (AFLP)*

AFLPs are generated from restriction digests of genomic DNA with enzymes that recognize both frequent and rare sites. This generates a large number of fragments, to which adapters of known sequences are joined (ligated), and then multiplied for several PCR cycles using non-selective primers (pre-amplification). Selective PCR amplification is then carried out with primers that recognize the adapter sequence plus 1-3 random bases. The amplified fragments are then separated electrophoretically and recognized through the use of radioactive or fluorescent labelling.

Like RAPDs, polymorphism is established through the absence of a fragment, which results in an essentially dominant marker system. The more random bases, the fewer the bands. There is therefore a trade-off between the ease of recognizing unique bans on the gel and the amount of data generated. Nevertheless, each primer combination generates a large number of bands, 10-20 of which may, on average, be polymorphic. AFLPs are a much more robust and versatile marker system than RAPDs and their ability to generate a large number of polymorphic bands for each primer combination gives the system a high multiplex ratio.

The use of AFLPs therefore generates a large number of data points in a short time and greatly accelerated the development of genetic maps. Waugh et al. compared AFLP loci in three barley crosses and found that they mapped to the same chromosomal locations, so AFLP markers could be used as anchor loci within a species. Precise sizing of AFLP fragments was, however, required to establish identical loci across several crosses and this was not always possible or practical, limiting the usefulness of the system in comparative mapping.

### *Simple sequence repeats (SSRs)*

SSRs, or microsatellites, are short tandem repeats of mono-di-, tri- and tetra-nucleotides although more complex repeats have been detected. They are abundant in mammals and have been found in plants by searching sequence databases. They can also be found by sequencing random clones of genomic DNA. The main advantage of SSRs over RAPDs and AFLPs is that they are, like RFLPs, co-dominant markers and can therefore reliably identify heterozygotes. They have also proved to be multi-alleleic, with over 30 different alleles being detected by a single SSR primer pair in barley. This facet renders them highly informative so they have high polymorphic information content (PIC) values.

The main drawback of SSR motifs has been found to greatly increase the rate of SSR discovery but the process is still resource hungry. A number of groups have invested resources in developing SSRs and 230 are available for wheat and over 560 for barley. Due to their high PIC values, SSRs are proving to be excellent anchor loci for comparing various studies and are the current marker system of choice within a species. Primers for SSR loci developed for one species, however, usually fail to amplify a product in another species and, even if a product is produced, it cannot be relied upon as a homologous locus.

### *Derivative marker systems*

A number of marker systems have been developed that share features of the other PCR based systems but differ in a number of respects. Two methods based on simple sequence repeats are inter-simple sequence repeats (ISSR) and *Copia*-SSR, also termed retrotransposon-microsatellite amplified polymorphism (REMAP). Two other methods, which are based on retrotransposons, are sequence-specific amplified polymorphism (S-SAP) and inter-retrotransposon amplified polymorphism (IRAP). All four are PCR based and produce a number of polymorphic bands for each PCR reaction. Results published so far suggest that all have a higher multiplex ratio than AFLPs, e.g. ISSRs in rice, S-SAPs in barley, IRAP and REMAP in barley. All are dominant marker systems and are therefore less informative than SSRs and their use as anchor loci has not been evaluated, but they do seem to have some promise as an alternative to AFLPs.

***Single nucleotide polymorphism (SNPs)***

SNPs occur as a result of a single base substitution, deletion or addition in a sequence and have been detected at a rate as high as 1 in 26 base pairs in barley. The advantage of SNPs is that specific oligonucleotides can be designed to detect a particular polymorphism in a positive or negative fashion. One can then analyse a batch of material with a high throughput and automated assay. Such a system also offers the possibility of dispensing with gel-based detection methods, which are one of the limiting factors in market analysis. As more sequence data become available, particularly from the commitment of the public and private research communities to develop large libraries of the public and private research communities to develop large libraries of clones representing the expression of single genes (expressed sequence tags – ESTs), it is becoming more realistic to identify SNPs on a large scale in the small-grained cereals.

At present, there is no simple, robust way of detecting SNPs other than amplifying a target sequence for a test panel of genotype and comparing all the sequences to reveal polymorphisms. This therefore requires a heavy investment in DNA sequencing but is now feasible with the latest generation of sequencers. Although SNPs have been found in barley which are diagnostic for the *rym4* and *rym5* resistance genes to Barley Yellow Mosaic Virus. It is likely that a combination of three or more successive SNPs in a sequence will be of more value in diagnostics.

The genotypic constitution at a number of successive loci (haplotype) gives much more information than the genetic constitution at an individual locus as the possible combinations are much more than the two at a single SNP locus. It is therefore likely that SNPs will become the main marker system in diagnostics but their value in generating genetic maps in unknown as yet, although their deployment would mean that maps would be based on known function rather than largely anonymous markers.

**Genetic Maps**

Genetic maps show the location of genes on chromosomes and the distances between them so that one can estimate whether or not two genes are likely to segregate together or, if breakage of a linkage is desired, the population size necessary to ensure recombinants between two neighbouring loci. Genetic maps are constructed by assembling segregation data for a series of markers, which can be used to estimate the location of genes relative to each other. With morphological markers, segregation could usually only be studied for a few loci at a time.

The standard procedure is to compare the frequencies of the various parental and recombinant classes in a segregating population such as an F2 or, more recently, doubled haploids. If two loci are completely unrelated, then the observed frequencies should be within the sampling error of that expected for unlinked loci. If the observed frequency of the recombinant classes is significantly less than that expected, then the loci are located in the same region and the degree of linkage can be estimated from the observed frequencies of the various marker classes. Thus, placing a new marker gene on a genetic map was matter of crossing it to a series of other known marker loci and checking the segregation ratios to see if there was any evidence of linkage. This lead to the placement of 38 loci across seven linkage groups in barley by 1951 and nearly 80 loci were assigned to the seven barley chromosomes by 1962.

Methodologies were developed to combine segregation data from a series of separate experiment to produce more inclusive maps which, for barley, were published annually in the *Barley Genetics Newsletter*. Mapping in this way was very much a matter of trial and error and it was not always possible to locate commercially important major genes. For example, the *Hordeum laevigatum* meldew resistance gene (*MlLa*) was found in the spring barley cultivars Vada and Minerva and many of their derivatives but a comprehensive conventional linkage analysis failed to localize the gene to any one barley chromosome. The development of molecular maps soon enabled the gene to be located on chromosome 2H.

### Association of major genes

The assignment of some commercially important major genes to particular chromosomes helped to elucidate some information about the control of quantitative characters. By developing random inbred populations that segregated for an easily characterized major gene, it is possible to assess the whole population for a range of economically important characters.

The variation for each character can then be partitioned into that ascribable to differences between the alternative alleles at the major gene locus and the variation within groups. If the former is significant when tested against the latter then the region of the genome in which the major gene is located also plays a role in the genetic control character measured. This method was used to find that the *sdw1* and *ari-eGP* dwarfing genes in barley were associated with a number of other characters. Whilst this analysis gave an indication of the possible consequences of deploying specific major genes, it is essentially retrospective and cannot distinguish whether association are due to linkage or pleiotropy.

### Mapping software

The development of large arrays of molecular marker data meant that manual mapping methodology was a far too labour intensive. Computer software was therefore developed to reduce the amount of manual calculation. The principles behind mapping remain the same but were updated to examine a large number of marker combinations simultaneously. The most popular (by citation) mapping software is MAPMAKER, although the more recent JOINMAP has gained in popularity, again by citation.

GMENDEL has also been used and all three differ in their approach to map construction but essentially produce similar maps from a given set of data. All have a reasonable user interface, are well documented and each has routines to evaluate the quality of the input data and/or the maps produced. They can all cope with the various different types of populations generally used in mapping (F2, backcross (BC), recombinant inbred lines (RIL) and doubled haploid (DH) inbreeding species such as the small-grained cereals. JOINMAP has the added advantage of being able to develop a composite map from different mapping populations but with some common reference markers.

### Physical maps

Genetic maps provide an interpretation of the distribution of genes along chromosomes based purely upon recombination. Using suitable stains, banding patterns can be directly observed on barley chromosomes. Some of these bands differ between barley cultivars and can be scored as genetic markers in segregating populations. Linde-Laursen combined a banding pattern marker with other markers to demonstrate that the genetic map location of the band differed from its physical location. This reflects the tendency for recombination to occur at the ends of cereal chromosomes, rather than at random.

The development of fluorescent *in situ* hybridization (FISH) provides an opportunity to visualize loci directly upon chromosomes and therefore extend the comparison of physical and genetic maps. Laurie et al. used ribosomal RNA markers to demonstrate considerable differences between the physical and genetic maps of the long arm of barley chromosome 2H. Kunzel and Korzun found no physical markers in some 70 cM of the distal part of the long arm of barley chromosome 3H, indicating that the genetic map was considerably lengthened in comparison to the physical. Differences between genetic and physical maps have important implications for those studying the organization of the genome but the genetic map represents what it is possible to achieve through recombination in terms of segregation of markers and is therefore the most relevant measure for those studying the inter-relationships of characters.

### Published maps

In the early 1990s extensive maps were published for two barley populations, Proctor × Nudinka, and Igri × Franka. Later the North American Barley Genome Mapping project developed extensive maps for two barley crosses, Steptoe × Morex, and Harrington × TR 306. The three ancestral genomes of bread wheat necessitated the construction of three sets of maps, with many homologous loci that to two or three of the constituent genomes. The early use of RFLps in wheat mapping was to construct maps from different crosses for each linkage group. The first genome-wide maps of wheat crosses involved its wild relatives but, because of the problems of low levels of polymorphism detected by RFLPs, an RFLP-based map of an inter-varietal bread wheat cross took longer to develop. Similarly, the first oat RFLP map to be published was a cross between wild species with a map from cultivated oats being developed later.

In addition, JOINMAP has been used to construct composite maps of barley over four and seven mapped populations. RAPDs were mainly used to fill the maps, using known RFLP loci as points to anchor the data to chromosomes. Several barley maps including RAPD makers were published, e.g. Vogelsanger Gold × Alf and Blenheim × E224/3. Becker et al. first reported the use of AFLPs in barley mapping and Powell et al. and Qi et al. reported other major barley mapping efforts using AFLPs. A largely AFLP-based map of a bread wheat cross has also been developed and AFLPs have also been used in oat mapping. SSRs are now being incorporated into genetic maps of barley and maps of bread wheat based on SSRs have also been published.

The most efficient current mapping strategy is to combine a number of reference SSRs with AFLPs and/or the derivative markers noted above to fill in the gaps. All of the above maps are notable for their relative lack of non-molecular markers. In the barley Proctor × Nudinka map, a major gene mildew resistance at the *Mla* locus on chromosome 1H and the naked locus *n* on 7H is presented. The lack of non-molecular markers reflects the lack of major gene morphological variants in crosses between cultivars, be they of barley, oats or wheat.

More recently, attempts to incorporate morphological markers on barley molecular maps have been made through the Oregon Wolfe Barleys and Kleinhofs et al. have attempted to place some barley morphological markers on molecular maps. With the large and ever-increasing numbers of molecular markers available for mapping in barley, oats and wheat, markers are available that give comprehensive genome coverage of all species. The problem is now establishing an order amongst a huge range of data. It is important to remember that the statistical resolution of genetic distances is dependent upon the number of individuals in the population being measured, as can be seen for the linkage formulae given by Mather.

Whilst mapping software usually estimates orders of loci to tenths of a centi-Morgan, these are beyond the resolution of most of the populations being studied and thus the fine-map order of a population of say 100 backcross or doubled haploid individuals cannot be relied upon. To solve this presentational problems,

Kleinhofs et al. have developed a system of 'bins' at stratified intervals along the chromosome. This has the advantage of obviating the need to integrate and present maps with huge numbers of data-points on them.

## Characters

One of the biggest benefits of having genetic maps is being able to localize genetic control of quantitative traits to specific regions of the genome, which requires genotyping and phenotyping of a random population of lines from a cross. For the same reasons that it is difficult for a plant breeder to select for important quantitative characters such as yield and quality, it is desirable to carry out phenotypic evaluation over a number of different environments. Thus, the phenotyping component is not trivial and ideally requires a population of inbred lines, either recombinant inbred lines (RILs) from a selfing series or doubled haploids (DH).

Prior to the advent of molecular markers, manipulation of pairing in wheat genomes had been carried out to develop special chromosome stocks that carried substitutions of whole chromosomes or chromosome segments. Phenotypic analysis of these stocks enabled quantitative, and qualitative, variation to be firstly located to a chromosome and then to more specific regions. This elegant series of experiments, reviewed by Law et al., allowed much progress to be made in the genetic analysis of wheat. The development of the stocks was laborious and limited the range of germplasm that could be surveyed. The development of molecular maps meant that quantitative traits could studied not only in a wider range of germplasm but also in more species.

### QTL Detection

Having assembled a set of genotypic and phenotypic data for a cross, the simplest way of detecting QTL is, for each marker in the genotype test, to classify the phenotypic data for each character into the parental marker groups. The means of each marker group can then be compared and tested for significance by analysis of variance or regression. Edwards et al. detected QTL for a range of traits in maize using such an approach. The disadvantage of this approach is that it is laborious and only gives an association of marker loci with a character. In cases where neighbouring marker loci are far apart, the differences between the marker groups may not reach significance for a QTL of small effect and/ or the QTL may be some distance from the marker.

Interval mapping was developed to step along a genetic map at set intervals and, at each interval, test for the presence of a QTL. The test is based upon the phenotypic means of the marker classes and the distance between the markers. This approach gives a more precise location of QTLs and MAPMAKER/QTL software has been developed to automate the procedure This approach used maximum likelihood and could efficiently detect single QTL effects on a linkage group. In the presence of more than one QTL per linkage group, the method could either fail to detect any effect at all, if the loci from a parent were of opposite sign, or detect a 'ghost' QTL.

Haley and Knott presented a least squares approach to interval mapping that gave similar results to maximum likelihood but was simpler to apply, could be adapted to a range of situations and was robust enough to detect multiple QTL in a linkage group. QTLs of large effect could mask others of small effect and Jansen proposed the use of co-factors to account for variation in other regions of the genome when scanning a target region. Jansen and Stam proposed using the whole marker set as initial co-factors and then eliminating them in a backward elimination regression procedure but this can lead to over-parameterization. Hackett proposed a forward selection method to identify co-factors and this has been adopted by Utz and Melchinger. Similarly Zeng proposed the use of marker co-factors to account for the effect of one QTL when searching for another.

The use of marker co-factors has been termed compound interval mapping and has been found, both by simulation and application to real data, to be more efficient at detecting QTL and also increases the precision of location. Software packages which implement this approach are MapQTL, QTL Cartographer and PLABQTL and MQTL can also analyse data for a trait from more than one environment and determine which QTLs are consistent enough over environments to be termed main effect QTLs and which are effective in one or more environments (QTL × environment interactions). PLABQTL differs from MQTL in its approach to this problem as it looks for QTL in the overall means from different experiments and then attempts to fit this model to each environment in turn. This means that, as MQTL searches the accumulated data over all environments for QTL, it is more likely to detect QTL × environment interactions of the cross-over type in addition to interactions of magnitude. Compound interval mapping is now the method of choice for QTL detection and choice of programme depends on objectives, availability and ease of use and interpretation. Marker regression has also been proposed as a method of QTL detection. This approach uses all markers in a linkage group to detect the presence of possible QTL and deviations of the observed values from predicted could indicate the presence of more than one QTL in a linkage group. This is a simple and easy to implement approach which has value in some situations but it does not detect QTL × environment interactions.

## QTLs Reported

A number of QTL studies in barley, oats and wheat have been published, largely centered around heading, height, yield and disease resistance. Some studies looking at malting quality parameters in barley and processing quality characters in wheat have also been published. Few comprehensive studies of agriculturally important traits have been published as most concentrate upon one or a few traits at a time. Barley appears to be an exception in that crosses have been studied for a range of agronomic, disease, yield and quality characters in North American, European and Australian germplasm. No comparable studies have been published in either oats or wheat as most of the results in the public domain have concentrated on one or a few characters. For example, separate studies in wheat have focused on QTLs for plant height, disease resistance, flour viscosity and frost tolerance.

There are fewer still published reports for oat but QTLs for resistance to Barley Yellow Dwarf Virus and groat oil content have been detected. In a number of the QTL studies conducted in barley, a major gene with a marked agronomic effect has been segregating, either a dwarfing or a disease resistance gene. In many cases, the major genes have been associated with QTLs for other characters. For example, the *sdw1* dwarfing gene on chromosome 3H in spring barley has been associated with malting quality characteristics. There are also a number of instances where QTL 'hot-spots' have been detected. For instance, the distal region of the short arm of chromosome 5H in Harrington × TR 306 is associated with QTL for yield, maturity and some malting parameters amongst other characters. The confidence intervals of

QTLs are so large that it is still not possible to determine whether associations with major genes or 'hot-spots' are due to genuine pleiotropy or to tight linkage. The construction of special genetic stocks in which a series of isolines differ by small, neighbouring fragments of the genome should help to resolve this question.

### QTL Validation

Before utilizing QTL information in a breeding programme, some form of verification of effect is desirable. This problem has been addressed in several practical studies in barley. Hayes et al. identified two QTL which, when combined, accounted for over 25% of the phenotypic variation in malt extract and some other malting quality parameters in the Steptoe × Morex mapping population. Han et al. found that, in another sample of lines from Steptoe × Morex, the two QTL accounted for much less of phenotypic variation for the characters.

Romagosa et al. came to the same conclusion in a study of yield in Steptoe × Morex. Beavis has carried out extensive studies of QTL effects in maize and concluded that accurate estimation of QTL effect requires extremely large populations. Working with populations of 100-200 lines resulted in considerable bias in estimate of effect. Bias in estimation of QTL effect appears to be a genuine problem but need not detract from results already accumulated, provided that the methodology identifies the most important QTL.

## Deployment of Molecular Markers

### Varietal Identification

Ainsworth and Sharp proposed that RFLPs could be used in varietal identification but, as noted above, the overall level of polymorphism detected by RFLPs is relatively low and the technique is demanding. Terzi demonstrated that RAPDs could also be used to distinguish between varieties of barley, oats and what but the disadvantage of this method is that the portions of the genome being sampled are unknown. Due to their high multiplex ratio, AFLPs are another possibility for cereal varietal discrimination. The dominant nature and low polymorphic information content of AFLPs may, however, make them less useful in distinguishing between closely related cultivars and, like RAPDs, the genomic distribution of the markers is likely to be unknown. SSRs appear to be particularly useful for varietal discrimination, due to their multi-allelic nature and high PIC values, and can provide unique genetic fingerprints of crop cultivars. Russell et al. used a panel of 11 SSR to distinguish between 24 winter and spring barleys either on the UK National or Recommended Lists. In fact three different subsets of four SSRs were shown to be capable of differentiating between all 24 barleys. It should be noted that two of the winter barleys were derived by the same breeder from the same cross and were morphologically similar but differed at 4 of the 11 SSR loci surveyed.

## Germplasm analysis

### *Cultivated germplasm*

Molecular markers have a number of attractions for determining the genetic variability in a germplasm pool and, with the development of genetic maps, the level of polymorphism detected can be related to genomic regions. A multi-variate technique called 'principal co-ordinate analysis can be applied to data collected from a panel of genotypes to reveal the pattern of relationships between them. This relationship can be inspected graphically either by plotting the scores for the first two or three principal co-ordinates or, if they fail to account for a sufficiently large portion of the variation, a dendrogram.

For barley, RFLPs have been used to differentiate between elite winter and spring European cultivars. Within these groups, discrimination between either two-and six-rowed types is possible amongst the winter cultivars and between feed and malting cultivars amongst the spring cultivars. RAPDs and AFLPs have also been used to examine the relationship between barley cultivars with similar results to those obtained from RFLPs. As for vertical discrimination, the disadvantages of using RAPDs and AFLPs is not knowing the chromosomal location of the markers.

SSRs have a number of advantages in studying genetic relationships between genotypes as the multiple alleles that can be detected at any one locus give a better overall impression of the variation. Amongst European spring barley, Russell et al. have shown that over 70% of all the alleles detected by a panel of 28 SSR loci distributed across all seven barley chromosomes were found in 17 'founder' genotypes which were largely old land-races, their early derivatives or disease resistance donors. Furthermore, more modern genotypes, as encompassed by a group released since 1985, possessed only 35% of the total alleles. This confirms the narrowing of the genetic base of modern barley, which Fischbeck suggested was due to:

1. The exploitation of relatively few land-races in plant breeding from the 1880s to 1920s;
2. A small number of cultivars that represented a significant advances and consequently featured heavily in subsequent breeding programmes; and
3. The use of exotic germplasm largely as a source of disease resistance.

Whilst this dependence on 'founder' genotypes is fairly universal across the genome, Russell et al. found that post-1985 spring barley cultivars did contain some novel alleles, notably in a region on chromosome 5H and they reflect selection for yield and/or quality.

Molecular analysis of wheat germplasm has not progressed as far as that of barley.. The relatively low level of polymorphism detected by RFLPs in wheat hindered progress and RAPDs were found to be unreliable due to primers amplifying different non-homologous sequences between genotypes. Paull et al. used a panel of mapped RFLPs to analyse a collection of Australian bread wheat genotypes to produce four major grouping that could be associated with the origin of the lines. They also found a number of differences between the observed and expected genotype based upon pedigree information. They concluded that, provided that the markers gave a broad, evenly-spaced genome coverage, the use of markers to estimate similarity was more meaningful than pedigree information as the latter did not account for selection and drift. Paul et al. used the low level of polymorphism detected by RFLPs to argue that, where differences existed, they might well reflect the effects of phenotypic selection.

In contrast, multivariate analysis of a sample of 11 winter wheat cultivars from Austria and Germany based on pedigree information, RFLPs, AFLPs and SSRs did not produce any meaningful groupings, nor were there any common trends between the four dendrograms. Over all the pairwise combinations of the 11 genotypes, there were no significant correlations between the pairwise combinations of genetic similarities estimated by three types of molecular markers and only the estimate produced by AFLP showed a significant correlation with pedigree. Working with oats, O'Donoughue et al. used principal co-ordinates analysis of

RFLP data scored on a panel of 84 cultivars. They found two major groupings from the analysis, which generally corresponded to winter and spring cultivars, and that some sub-groups could be associated with breeding history. Whilst they found a significant correlation between genetic distance, as measured by RFLPs, and pedigree, like the studies of wheat and barley noted above, the correlation was small.

### *Exotic germplasm*

ISSRs have been used not only to discriminate between winter and spring cultivated European barleys but also to highlight the distinctiveness of wild barley *Hordeum spontaneum*. This has been explored further with the use of mapped SSRs, which has highlighted genetic bottlenecks that have occurred at specific regions of the barley genome, notably on chromosome 7H. A survey of land-race material collected by ICARDA shows that it is intermediate between *Hordeum spontaneum* and cultivated barley in terms of alleleic diversity.

Greater diversity exists in the land-races than in *Hordeum spontaneum* in specific regions of the genome, much of it from novel alleles. In *Hordeum spontaneum* Pakniyat et al. analyzed single accessions from 39 collection sites with AFLP derived from 12 primer combinations. They found that the first three principal co-ordinates grouped the accessions according to the eco-geographic variables associated with the collection sites. The first principal co-ordinate differentiated between the sites on the basis of altitude whilst the second and third discriminated between longitude and latitude respectively. If one could then identify the phenotype that is favoured at a particular site, then such an analysis provides a powerful method of studying the direct action of natural selection.

### Pedigree Analysis

The multi-allelic nature of SSRs makes them extremely useful in analyzing pedigrees of genotypes. Swanston et al. found a barley SSR, Bmac213, was linked within 7 cM of a major gene locus affecting non-production of epi-heterodendrin (a cyanogenic glycoside precursor important in some distilleries). They were able to trace the origin of this allele back through the pedigree of the spring barley cultivar Derkado to the cultivar Emir. Other derivatives of Emir were also non-producers of epi-heterodendrin and also possessed the same Bmac213 allele as Derkado. The origin of the Emir allele was probably the disease resistance donor Arabische, although the accession tested neither possessed the Derkado Bmac213 allele, nor was it a non-producer of epi-heterodendrin. This was probably because the accession used in the breeding of Emir differed from that in the collection studied. This illustrates the general problem that material in collections, particularly less homogeneous material, may not accurately represent the variation in a particular genotype due to genetic drift and/or difficulties in ensuring that all the variation is passed on from generation to generation during multiplication and maintenance of accessions.

## Marker-assisted Selection

With the development of molecular maps, there are now many published association of characters with markers that are close enough to be used in marker-assisted selection programmes. Theoretical studies have highlighted the benefits of applying marker-assisted selection in breeding programmes of the small-grained cereals which suggest that marker-assisted selection should offer a substantial advantage over phenotypic selection for characters of low heritability under intense selection. The problem then becomes one of identifying robust QTLs that can be used in marker-assisted selection for such characters.

Various selection strategies, including marker-assisted selection, have been compared for malt extract and some other malting quality characters and yield in barley. Both studies demonstrated that marker-assisted selection was effective, particularly when combined with phenotypic selection in the later stages of a breeding programme. A strategy of using genotypic selection to assemble a pool of 'elite' germplasm in which a number of desirable QTLs of major effect for traits that are difficult to measure followed by phenotypic selection to identify the best of these lines has a number of advantages:

1. Selection is concentrated on lines that are most likely to meet the desired standard for yield and quality.
2. The number of lines in the advanced stages of breeding programmes is reduced.
3. Response from genotypic selection of QTL of small effect is, at best, likely to be negligible and therefore not cost-effective.

Marker-assisted selection in commercial breeding of wheat and barley is carried out, but generally for qualitative characters that are difficult to assess phenotypically. Markers can also be used to gain some knowledge about unknown germplasm, for example to choose parents for use in crossing. Alternatively, where progeny are known to be segregating for a qualitative character, marker-assisted selection may be applied at varying stages of the selection programme to identify those carrying the target gene.

The barley Yellow Mosaic Virus complex is one example where molecular markers have been deployed. The rym4 resistance gene on barley chromosome 3HL was found to be flanked by the RFLPs MWG10 and MWG838. The latter was converted into a sequence tagged site (STS) and Tuvesson et al. demonstrated its use in molecular-assisted selection. The SSR Bmac29 is also linked to the *rym4* locus and is capable of differentiating not only between resistant and susceptible alleles but also between the *rym4* and *rym5* alleles.

**Backcross Conversions**

The benefits of using marker-assisted selection in backcrossing have been demonstrated theoretically. Toojinda et al. identified a source of barley stripe rust (*Puccinia striiformis*) resistance in exotic germplasm and then used a backcross conversion scheme to introgress the resistance into the locally adapted cultivar Steptoe. Progeny from the first backcross that carried the desired genotype were selected by using RFLP markers that were polymorphic and flanked QTL. The selected BC1F1 genotypes will be heterozygous for the target segment and therefore doubled haploids were produced from them to produce inbred lines. The doubled haploids were then re-selected with the flanking RFLP markers to identify those that carried the QTL. AFLPs were used to screen the background genotype and showed that he percentage of the donor genome varied from 7 to 60%. Under a normal backcrossing scheme, the percentage of the donor genome is expected to average 25% in a first backcross so some of the lines approached the equivalent of a third backcross.

The capacity to identify such lines demonstrates the power of marker-assisted selection in a targeted backcrossing strategy. Selected backcross inbred lines can be produced within a two-year time-scale, regardless of whether the crop is winter- or spring-sown. The scheme is most suitable for one or two loci and, because it relies on selection of heterozygotes at the marker loci, works best with co-dominant markers for the target locus. If working with exotic germplasm, the relatively low levels of polymorphism detected by RFLPs should not be a problem. Given the increasing numbers of loci that cover the genomes of barley and wheat and the other advantages noted above, SSRs are the current marker of choice in targeted backcrossing schemes.

## FUTURE PROSPECTS

### Candidate Genes

Expressed sequence tags (ESTs) are developed by partially sequencing cDNA clones from target tissues. From the sequence information collected, searches can be made against publicly available databases and, given an adequate level sequence homology, some idea of function can be derived. A big international effort is being made to assemble comprehensive EST databases for the *Triticeae* that will be publicly available. At the same time, private companies are investing heavily in developing EST libraries for their 'core crops but it is unlikely that much of their information will be made public.

The challenge in both the private and public sector is to find ways of incorporating ESTs on genetic maps in order to determine whether any correspond to QTLs of interest, which would facilitate the change in emphasis in genetic maps from largely anonymous to known-function markers. Combining this information with QTL mapping will not only enable some functionality to be assigned to a particular QTL but also provide sequence data for its direct manipulation. Current QTL mapping methodology is, however, far too imprecise to enable such assignations to be made unambiguously. Mapping large EST libraries also poses a number of problems, including choice of the most appropriate marker system. Some 5% of barley EST sequences have been found to possess SSRs but their informativeness has yet to be established. The identification of SNPs in ESTs appears to be more informative but identification currently requires a massive sequencing effort to reveal differences in panels of genotypes.

### DNA Chips

DNA sequences can now be assembled in micro-arrays over a small area—so called DNA chips. DNA chips are being advocated as suitable for a number of applications, including genotyping where they would be highly suitable for the detection of bi-allelic markers such as SNPs. The high level of automation that is possible with DNA chips means that it would be possible to carry out high through-put genotyping, which would enable marker-assisted selection on a large scale.

DNA chip technology is still under development and its cost is not yet apparent. Whilst it appears promising, it will have to show a clear benefit over current approaches before it becomes a worthwhile investment purely for germplasm screening purposes. Apart from some obvious qualitative characters, the problem of identifying QTLs that are robust enough to warrant application of marker-assisted selection will limit the amount of screening that could currently be done using DNA chips. If the technology was sufficiently cost-effective, then it may, however, be suitable for screening QTLs that are less robust.

**Bioinformatics**

The development of SSRs for a wide range of crop species has led to the possibility of not only examining genetical relationships through principal co-ordinates analysis but also using the map locations of the markers to produce a more meaningful graphical representation of the allelic constitution of genotypes. Computer software has been developed to construct these graphical representations, e.g. Supergene, but no package published so far can cope with the range of alleles that can be generated at one locus by SSRs.

As SSRs are integrated into mapping studies of qualitative and quantitative characters, it becomes possible to assign values to specific SSR alleles. A variety of crossing strategies could then become possible by genotyping parental lines. Working within adapted germplasm, one could:

1. Assemble all the favourable QTL alleles for important characters that are difficult to screen for in cross combinations or
2. Make crosses between parents that both possess the major favourable QTL alleles and use conventional phenotypic selection to assemble the minor QTL alleles in a favourable combination.

Combining the information in such a database with genotypic information from the unadapted gene-pool would facilitate the identification of novel alleles in the region of important QTLs. These novel alleles may also represent novel variation for the character and can be introgressed through a marker-assisted backcrossing scheme to test whether or not they represent useful variation in wild relatives has already been detected in tomato and rice, although it remains to be thoroughly tested whether the QTLs from wild relatives represent new loci or variants at existing loci. The approach presented by Tanksley et al. is interesting as it combines mapping with introgression in a methodology termed 'advanced backcrossing'. Its utility in the small-grained cereals discussed in this chapter remains to be evaluated.

### Recombinant Chromosome Substitution Lines (RCSLs)

One of the problems in establishing reliable QTL locations is that the estimate is usually based on very few recombinant lines and therefore it is not surprising that QTL effects vary in different studies, even when sampling from the same cross. One solution to this problem is to create a series of isogenic lines which differ only a small (ideally contiguous) segment of chromosome from a donor line. These can then be replicated and widely tested along with the recipient cultivar to estimate the effect of individual segments in isolation. This approach can be applied in sequential manner by an initial screen of lines with relatively large introgressed segments to identify candidates for further rounds of backcrossing to break up a segment into a series of smaller ones. This approach can therefore be used to derive more accurate estimate of QTL effect and position to refine the target segment to be used in market-assisted selection. Another benefit of this approach is that it should help to distinguish between close linkage and pleiotropy, as well as providing material for gene cloning.

## CONCLUSIONS

Molecular markers have undoubtedly revolutionized the genetic analysis of performance traits and can be applied in marker-assisted selection schemes. Markers have been successfully applied in plant breeding but there are few published examples. Those that have been published are largely concerned with the manipulation of qualitative characters that give a characteristic phenotype. The identification of QTL influencing important quantitative characters does not yet seem reliable enough to warrant the use of markers tagging QTLs in a marker-assisted selection programme. What is certain is that the technological developments will continue and that much more information about the genetics, biochemistry and physiology of crop plants will emerge. The challenge is, and will continue to be, to integrate all the information with conventional phenotypic selection into efficient plant breeding programmes.

# Chapter 3

# Biotechnology in Agriculture

Technological advances biotechnology, including genetic engineering, have enabled transfer of genetic traits both within species and between entirely different plant and animal species. Currently, biotechnology techniques are being used in various fields, including agriculture, veterinary medicine, pharmaceutical development, forestry, energy conservation, and waste treatment. These techniques, if applied responsibly, have the potential to increase productivity in crops and livestock, control pests, produce new foods and fiber crops, and develop effective medicines. Potential environmental and economic benefits from biotechnology include the reduction of fossil fuel in agriculture and forestry through improved nutrient availability in crops and livestock, use of fewer artificial inputs (e.g., synthetic nitrogen fertilizers, insecticides, and fungicides), and more cost-effective and environmentally friendly waste management practices, such as bioremediation. If realized, these improvements will help protect ecological system by reducing habitat degradation. In addition, some of the biotechnology techniques should improve the economics of agricultural and forestry production systems. Although genetic engineering can be expected to provide major benefits to agriculture and the environment, risks with the use of this technology should also be recognized.

## Disease-resistant Crops

### Engineering Virus-resistance in Crops

Resistance against crop disease in plants, caused by viruses, bacteria, and fungi is now being explored through biotechnology and genetic engineering techniques as a way to reduce the loss of crops. Because viruses in the field cannot easily be treated, the production of genetically engineered, virus-resistance crops is agronomically significant. In addition, few antibacterial chemicals are available to control bacterial diseases. It has been estimated that viruses, bacteria, and fungi are collectively responsible for significant crop losses estimated at 12%, or nine hundred million tons, of preharvest yield worldwide. More than 350 field tests of genetically engineered disease-resistant plants have been approved in the United States since 1987, and the majority of these have been created to produce disease-resistant, genetically-engineered crops impervious to viral infections. Success in engineering virus resistance in tobacco, alfalfa, potato, cucumber (*Cucumis sativus*), melon (*Cucumis melo*), alfalfa, and tomato plants have been reported by Cuozzo et al. (1988), Hill et al. (1991), Truve et al. (1993), Gonsalves et al. (1992), Dong et al. (1991), and Xuē et al. (1994), respectively.

Field trials with tobacco containing the gene form the mosaic virus for the production of the coat protein have shown that resistance can be transgenically induced. For example, in China, field trials of tobacco that contains the tobacco mosaic virus and tomatoes with cucumber mosaic virus are under way. Efforts are also being aimed at rice because of its importance as a staple crop in this region. In Japan, a method for producing fertile transgenic rice plants using an electroporation system has been developed.

Transgenic rice plants expressing the rice stripe virus-coat protein (RSV–CP) have been developed to fight the rice stripe virus, one the major viruses of rice plants in Japan, Korea, China, and Taiwan. The findings of a 3-year biosafety study of ecological risks have demonstrated that expressing the introduced gene (RSV–CP) in a japonica rice variety (Kinuhikari) resulted in transgenic rice plants that did not: (1) affect morphological and ecological traits with the exception of some somaclonal variations, (2) hybridize with closely grown rice plants, (3) exhibit the tendency to become weeds, (4) produce any detectable toxic substances, and (5) have any observable effects on subsequent cultivation, microorganisms in soil, insects in florae, or on surrounding plants. However, these results will need to be followed up with longer-term biosafety assessment in the future. In the United States, squash and the papaya are two of the more recent models of crop engineering for virus resistance.

In 1994, the genetically engineered, virus-resistant squash developed by Asgrow seed company for resistance to zucchini yellow mosaic virus (ZYMV) and watermelon mosaic virus II (WMV II) was one of the first genetically engineered crops commercialized in the United States. Researchers have also developed two genetically engineered papaya lines by utilizing rDNA techniques to isolate and clone a papaya ringspot virus (PRV) that encodes for the production of the viral coat protein. Papaya, one of the three largest crops in Hawaii, has been decimated in recent years by PVR. Hawaiian papaya growers believe these two lines of genetically engineered, disease resistant papaya could save the $45 million Hawaiian papaya industry from extinction. However, it should be noted that, although the mild strain of PRV displayed excellent resistance to PRV isolates form Hawaii, it showed only moderate to no protection to isolates from different geographic regions.

**Engineering Bacterial and Fungi Resistance in Crops**

Harms (1992), who has reviewed recent developments in the production of resistance to fungal and bacterial diseases via genetic engineering points out that, 'although there has been much research on how to incorporate such resistance into crop plants, few improvements have been made in this area. Chen and Gu (1993) have described efforts that are being made to combat bacterial blight, which can reduce rice yields by as much as 10%, through genetic engineering. Developing disease-resistant crops should also receive high priority secondary to the large amounts of fungicides that are currently applied to fruit and vegetable crops. Aspelin et al. (1994) reported that in 1993 131 million pounds of pesticidal active ingredient was applied at a cost of $584 million. Fungicides are sometimes harmful to beneficial insects and toxic to earthworms and many other beneficial soil biota.

The number and activity of these soil biota are important in maintaining soil fertility over time because they recycle nutrients in organic matter and aid in water percolation and soil aeration. Furthermore, fungicides rank highest for carcinogenicity of all pesticides applied to agriculture and account for approximately 70% of human health problems associated with pesticide use. One way to reduce crop losses to fungi and the external application of fungicides is to introduce genes that encode proteins with antifungal properties into crop plants. Several genes have been identified so far in fungi, bacteria, and plants that are effective for the engineering of resistance to fungi based on their ability to produce enzymes, such as chitinase, that attack the cell wall of fungi.

Transgenic tobacco plants with a chitinase gene from beans produced elevated levels of chitinase in roots and leaves compared with control plants in greenhouse experiments. Both experimental and control plants were grown in soil inoculated with the fungal pathogen *Rhizoctonia solani*. A positive association was also found between the level of chitinase expressed in the experimental plants and survival. Broglie et al. (1993) and Lin et al. (1995) also reported some success has been achieved with engineering resistance to the stem out pathogen (*Rhizoctonia solani*) in oilseed rape or canola (*Brassica napus*) and rice, respectively. The engineering of resistance to the fungus *Fusarium oxysporum* in the tomato has also been successful.

### *Risks*

#### *Creation of new weeds*

In terms of risks, it has been proposed that large-scale cultivation of plants expressing viral and bacterial genes could lead to novel ecological risks. The most significant ecological risk would be gene transfer via pollination from cultivated crops to wild relatives. For example, it has been postulated that the virus-resistant squash commercialized in 1994 could transfer its newly acquired virus-resistance genes to wild squash (*Cucurbita pepo*), which is native to the southern United States, where it is an agricultural weed. If the virus-resistance genes were to spread, newly disease-resistant wild squash could become a hardier, more abundant weed. Moreover, because the United States is the origin for squash, changes in the genetic make-up of wild squash could lessen its value to squash breeders. Another area of concern is the production of virus-resistant sugar beets, which is likely to result in exchange of genes between cultivated and wild populations of beets (*Beta vulgaris* L.) because production areas contain wild or weed beet populations, or both, separated by only a few kilometers. A genetic exchange could take place owing to wind pollination, biotic pollination, or the common gynoniecy of wild beets. A genetic introgression from seed beet to wild beet populations had already been observed in Europe.

*Virus that infect new hosts*

Some plants pathologists also hypothesize that development of virus-resistant crops may allow viruses to infect new hosts through transencapsidation. This may be especially important for certain viruses (e.g., luteoviruses) for which possible heterologous encapsidation of other viral RNAs with the expressed coat protein is known to occur naturally. With other viruses, such as the PRV, risk of heteroencapsidation is thought be minimal because the papaya itself is infected by very few viruses.

*Creation of new viruses*

Virus-resistant crops may also lead to the creation of new viruses through an exchange of genetic material or recombination between RNA virus genomes. Recombination between RNA virus genomes requires infection of the same host cell with two or more viruses. Several authors have pointed out that recombination may also occur in genetically engineered plants expressing viral sequences on infection with a single-virus and that large-scale cultivation of such plants may lead to increased possibilities of combinations.

It has recently been shown that an RNA transcribed from a transgene can recombine with an infecting virus to produce highly virulent new viruses. An overall strategy of risk assessment utilizing an incremental approach entails: (1) identifying potential hazards, (2) determining frequency of recombination between homologous but nonidentical sequences, and (3) determining whether such recombinants can have selective advantage.

Fernandez-Cuartero et al. (1994) have already demonstrated that, even tough a particular pseudo-recombinant strain was at a competitive disadvantage relative to a parent cucumber mosaic virus (CMV) strain, a spontaneous recombinant that arose from the pseudorecombinant (resulting from pseudorecombination of the situation in which gene components of one virus are exchanged with the proteins of another coat) had enhanced fitness relative to either of the other original strains.

## Herbicide-Resistant Crops (HRCs)

At the moment at least 4 engineered crops for target herbicide resistance are on the market, and 13 among the key crops in world production have been extensively tested in field trials. In addition, some crops (e.g., corn) are being engineered to contain both herbicide (Glyphosate) And insecticide-resistance (Bt δ-endotoxin) (Gene Exchange 1997). The potential benefits and risks of herbicide-resistant crops (HRCs) are discussed in this section.

## Potential Benefits Associated with the Use of HRCs

### *Possible reduced use of herbicides*

Proponents have argued that reduction of herbicides adopted for HRC crops occurs primarily because these "new" herbicides are needed in lower does (if compared for instance with atrazine, 2,4-D, and alachlor) and are applied later in crops, post-emergence. However, higher resistance of the crop to the target herbicide would, in practice, suggest to the farmer to adopt a higher rate than advised to ensure that all weeds are burned in one tractor trip with the targeted broad spectrum herbicide.

### *Improved integrated pest management (IPM)*

Integrated pest management IPM could benefit from some HRCs if alternative nonchemical methods were applied first to control weeds and the target herbicide were used later, only when and where the threshold of weeds is surpassed, in postemergence. If HRCs could be implemented in IPM programs without underestimating the induction of weed resistance and adopting all the available nonchemical alternatives to manage weed control, this technology would be a step toward more sustainable agriculture. However, in practice, insufficient work of extension outreach and appropriate protocols promoted by the producers could only lead to a further link of the farmland to the producers and their marketing policies aimed at increasing their sales of the targeted herbicides aside to their HRCs seeds.

### *Benefit to developing countries*

Although the majority of HRCs currently on the market and under development belong to a key crops in Western agriculture, a few innovations have been proposed that would help developing countries. For example, HRCs have been proposed for improved control of parasitic flowering seeds such as *Orobanche* and *Stringa*, both which can severely reduce grain yields. The HRCs would permit more effective herbicide action against the soil parasitic weed without damaging the target crop. Trials on boomrape have demonstrated that the engineered plants can overproduce at a rate at least double the reproduction of the control plants. However, the authors observed that this technology can only be used with weeds that do not have the potential to interbreeds with wild relatives that could themselves become weeds. For example, in northern African countries, most crops such as sorghum, wheat, and canola (oilseed rape) have their wild relatives nearby, which therefore increases the risk that genes from the herbicide-resistant crop varieties can be transferred to wild relatives. The same gene escape risks are possible for tomato, corn, and potato in South America.

## Potential Risks Associated with Use of HRCs

### *Building weed resistance*

The risk that herbicide-resistant genes from a transgenic crop variety can be transferred via pollination into weedy relatives has been demonstrated for canola (oilseed rape) and sugar beet. Mikkelsen et al. (1996) and Brown and Brown (1996) have shown that herbicide-resistant genes from a resistant genes from a transgenic canola move quickly into wild relative weedy populations. Boudry et al. (1994) also noted consistent gene flow between the cultivated sugar beets and weed beet populations. Repeated use of the same herbicide in the same area creates problems of plant resistance to the target herbicide. This concern has consistent bases in the recent history of herbicides.

For instance, if glyphosate from an actual few million hectares of crops were allowed to associate with HRC crops, the resulting acreage could be around 70 million ha, and pressure on weeds to evolve resistant biotypes could be pronounced. Sulfonylureas and imidazolinones, to be targeted in HRC crops, are particularly prone to rapid evolution of resistant weeds and have already resulted in several cases of resistance. Extensive adoption of HRCs will increase the acreage and surface treated, thereby exacerbating the resistance problems.

### *Environmental risks*

Even if less environmentally persistent than previous herbicides (e.g. Alachor, 2,4-D atrazine), the "environmentally friendly" HRC have, as do most chemical pesticides, consistent or severe environmental effects.

#### *Bromoxynil*

Bromoxynil has been targeted in HRC cotton by Calgene and Monsanto. This herbicide has traditionally been used in winter cereals, cotton, corn, sugarbeets, and onions to control large-leaf weeds. Drift has been observed that has resulted in damage to nearby grapes, cherries, alfalfa, and roses. In addition, leguminous plants can be very sensitive to this herbicide, and potatoes can be damaged as well. Consistent residues over the accepted standards have been detected in soil and groundwater and as fallout. Rodents tested have demonstrated some mutagenic responses. Stafilinid beetles have been shown to have reduced survival and egg production at suggested dosages. Crustaceans (*Daphnia magna*) have also been severely affected.

*Glufosinate/Bialaphos*

Many crops have been modified for this herbicide-resistance PAT (phosphinothricin acetyl transferase) gene, which has been introduced into alfalfa, corn, barley, wheat, rice, canola, peanuts, soybeans, sorghum, tomatoes, and sugarbeets. Turf-grass (*Agrostisis stolonifera*) and other components have been engineered for resistance but need appropriate environmental risk assessment before being marketed. This herbicide has been on the market since 1984 as a synthetic development of a natural pathogen toxin or *Streptomyces viridochromogenes*. The amount of active ingredient used is 200-900 g/ha postemergence on the engineered plants. Although detrimental effects on users and consumers seem unlikely under recommended doses, toxic effects on humans and animals have been reported. For example, the Basta surfactant (sodium polyoxyethylene alkylether sulfate) has been shown to have strong vasodilative effects in humans and cardiostimulative effects in rats. Treated mice embryos exhibited specific morphological defects.

*Imidiazolinone*

Imidiazolinone is applied at low doses (38-50 g/ha) in beans and soybeans postmergence. It has been observed that, at the field rate of 50 g/ha, there is no effect in laboratory and field microbial biomass. However, higher doses induce catalase and hydrolase activity and increases the risk for monocultural practices and reduced mycelial growth in *Sclerotinia trifoliorum*.

*Sulfonylurea*

Sulfonylurea is used as a herbicide on wheat, barley, sugarbeets, cotton, maize, potatoes, and soybeans postmergence. Drift from very low amounts (5-30 g/ha) can damage cultures, and potential losses may result in several crops, wild plants, and nontarget invertebrates.

*Glyphosate*

Most HRCs have been engineered for glyphosate resistance. Although adverse effects of herbicide-resistant soybeans have not been observed on feeding animals such as cows, chickens, and catfish; genotoxic effects have been demonstrated in other nontarget organisms. Earthworms have been shown to severely damaged by the glyphosate herbicide at 2.5-10 1/ha. For example, *Allolobophora caliginosa*, the most common earthworm in European, North American, and New Zealand fields is damaged by this herbicide. In addition, aquatic organisms, including fish, can sometimes be severely damaged. The prevalence of the nematode *Steinerema feltiae*, a useful biological control organism, is reduced by 19-30% by use of this herbicide.

***Health risks***

The unknown health risks associated with the use of herbicides (as well as most xenobiotics) involve the effects of low-level chronic exposures. Although most research has addressed cancer risk, much less research has been focused on neurological, immunological, developmental, and reproductive effects. Much of this problem reflects the dearth of methodologies and diagnostic tests at the disposal of scientists necessary to evaluate the risks caused by exposure to many chemicals, including herbicides, properly. Although industry often stresses the desirable characteristics of their HRCs, environmental and alternative agriculture groups, as well as some scientists, disagree, with the contention that these products are safe. For example, research has shown that the application of glyphosate can increase the level of plant estrogen in the bean *Vicia faba*. Feeding experiments have shown that cows fed transgenic glyphosate-resistant soybeans had a statistically significant difference in daily milk fat production compared with other test groups.

**Increased Costs Associated with Use of HRCs**

Because the herbicides for which HRCs are being designed are almost all under patent, they will be more expensive than many of the herbicides they are intended to replace. In addition, although analysts project that, for example, switching to bromoxynil for broadleaf weed control in cotton could result in savings of $37 million each year from reductions in herbicide purchases of 40% to 50%, few economic product evaluations that demonstrate cost savings with the use of HRCs have been published.

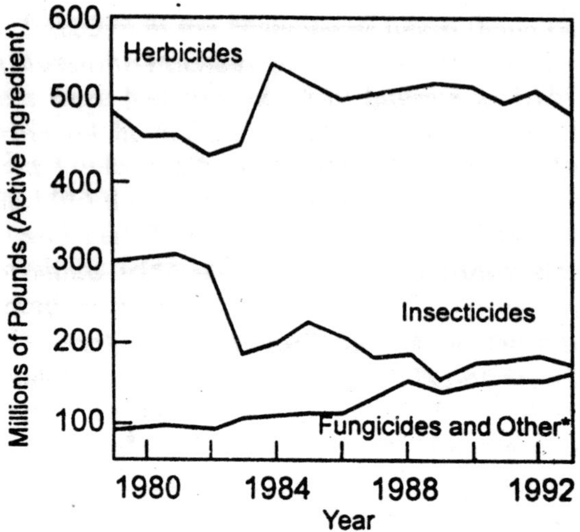


* Other = rodenticides, fumigants and molluscicides

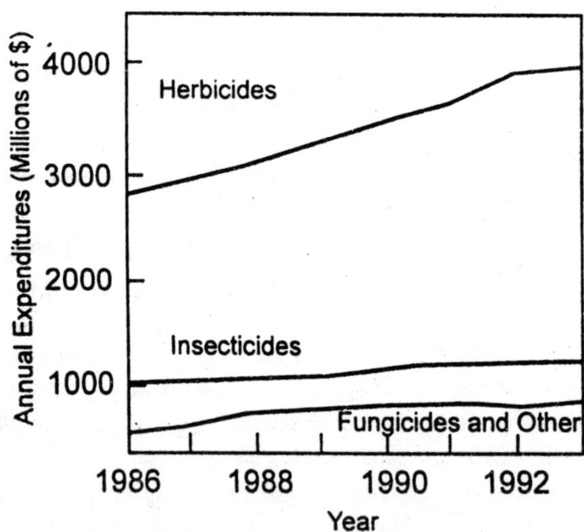


*Fig. 3.1. The choice of herbicide resistance as the main target for engineered crops seems related to the promising trend of herbicide seals and returns rather than to other environmental strategies such as potential reduction of pesticides in the environment.*

Furthermore, recent problems with use of glyphosate-resistant cotton in the Mississippi Delta region (cross losses resulting in up to $500 thousand of this year's cotton crop) suggest this technology needs to be perfected further before some farmers will reap economic benefits. Some scientists suggest that use of HRCs will cause a shift to the use of fewer broad-spectrum herbicides, which will reduce the amount of sprays and the amount of herbicide used per application, others contend that the overall use of HRCs will actually increase herbicide use and thereby increase costs associated with their application.

## Bacillus thuringiensis (BT) for Insect Control

More than 40 BT crystal protein genes have been sequenced and 14 distinct genes identified classified into 6 major groups based on amino acids and insecticidal activity. Many crop plants have been engineered with the BT δ-endotoxin, including alfalfa, corn, cotton, potatoes, rice, tomatoes, and tobacco. The amount of toxic protein expressed inside the modified plant is 0.01-0.02% of the total soluble proteins.

### *Potential economic and environmental benefits*

#### *Corn*

Supporters of BT usage cite current trials demonstrating a high level of efficacy in controlling corn borer damage on plants. Corn engineered with BT δ-endotoxin has the potential to recover 5-15% of corn borer damage in 70 million acres in the United States with a projected benefit of $50 million annually. However, without careful management tactics based on university agriculture extension and technical expertise, growers and seed suppliers could see unforeseen environmental risks.

#### *Cotton*

Cotton was the first crop plant engineered with the BT δ-endotoxin released into the market. It has been estimated that caterpillar pests, including the cotton bollworm and budworm, cost U.S. farmers about $71 million/year as measured in yield losses and insecticide costs. Benedict et al. (1992) predicted that the widespread use of BT-cotton could reduce insectide use and thereby decrease costs by as much as 50 to 90%, which would save farmers $86 to $186 million/year, respectively.

#### *Potatoes*

Genetically engineered Russet Burbank potatoes with *Bacillus thuringiensis tenebrionis* (the δ-endotoxin is represented inside the engineered potatoes at 0.05-0.1% CryIIIA as a percentage of total protein) have shown to be successful. This formulation is also suggested as a base to develop an effective and sustainable IPM program for potatoes.

*Eggplants*

The δ-endotoxin Cry3b-engineered eggplants have demonstrated consistent effects against potato beetles. However, resistance is one harrowing prospect, and proper resistance management has to be considered.

***Potential environmental risks***

Insects that develop resistance to transgenic crop varieties are one of the possible risks associated with the use of the BT δ-endotoxin in genetically engineered crop varieties. Resistance to BT has already been demonstrated in the cotton budworm and bollworm. If BT-engineered plants become resistant, a key insecticide that has been utilized successfully in IPM programs, could be lost. Therefore, proper resistance-management strategies with the use of this new technology are imperative. Another potential risk is that the BT δ-endotoxin could be harmful to nontarget organisms. For example, it is not clear what potential effect BT δ-endotoxin residues, that are incorporated into soils will have against an array of nontarget useful invertebrates living in the rural landscape.

## Recycling of Toxic Wastes and Improved Waste Management

Advances in technology have resulted in the availability of a variety of chemicals, many of which have increased ecosystem pollution. Currently, in the United States, some 70,000 different chemicals are released into the environment through soil, water, and air; an estimated 100,000 chemicals are used worldwide. Cleanup of hazardous wastes by conventional technologies is projected to cost between $400 and $750 billion in the United States alone on the basis of estimates obtained from a variety of federal and private sources. Various strategies that are utilizing genetic engineering and biotechnological methods to deal more efficiently with waste removal and management include bioremediation, phytoremediation, and the production of biodegradable plastics.

**Bioremediation**

Biotechnology and genetic engineering may help reduce environmental pollutions through bioremediations, which is the application of biological treatment using microbes to degrade chemical materials at polluted sites effectively. Because bioremediation provides continuous cleanup of contaminated sites, such as pesticide residues in agricultural ecosystems, it has significant advantages over other techniques. Furthermore, a marked degree of self-regulation is present in such systems because the added microbes survive by consuming and degrading chemicals but die off when the nutrient source is reduced or eliminated.

Investigations into genetically modified bacteria for bioremediation have produced a strain of *Pseudomonas cepacia* that degrades a wide spectrum of chemical pollutants, including vinyl chloride, dichloroethylene, phenol, toluene, xylene, and creosol. Most importantly, this bacterium produces an enzyme that is capable of degrading trichlorethylene, a persistent industrial degreaser. Bioremediation techniques are growing in importance and diversity as new approaches are developed to use both wild and genetically engineered biota for chemical pollution abatement.

**Phytoremediation**

Phytoremediation, or use of specially selected and engineered metal-accumulating plants for environmental cleanup, is an emerging technology that may serve as a cost-effective approach to treating soils and groundwater contaminated with toxic metals. Currently, cleanup of sits in the United States contaminated with heavy metals and organic mixtures can alone cost $42.5 billion. Phytoextraction, or the use of metal accumulating in plants to remove toxic metals from soils, could be utilized to clean up sandy loam soil to a depth of 50 cm, which would cost $60,000-$100,000 compared with at least $400,000 for excavation and storage alone using traditional soil removal methods.

Rhizofiltration, or the use of plant roots to remove toxic metals from polluted waters, may offer an advantage in water treatment because of the ability of plants to remove 60% of their dry weight as toxic metals, thus markedly reducing the generation and disposal costs of hazardous radioactive residue. Rhizofiltration could also be a cost-competitive technology in the treatment of groundwater containing low concentrations of toxic metals. Genes encoding the cadmium-binding protein, methallothionein, have been recently shown to be expressed in plants. A research group at Peaking University has engineered a human gene encoding methallothionein (a protein that binds heavy metals) into tobacco. The genetically engineered plants have survived exposure to cadmium concentrations that were 25 times greater than the dose that killed control plants. More importantly, these genetically engineered plants have also been shown to absorb cadmium from the soil. Researchers are also now engineering the gene into weeds in the hope of using the transgenic weeds to reduce heavy metal contamination on farmland.

### Biodegradable Plastics

Another area of waste management biotechnology and genetic engineering could play a role is in the production of biodegradable plastics. Currently, plastics account for 20% by volume of all municipal solid wastes. Estimates of the current global market for biodegradable plastics range up to 1.3 billion kg per year. However, until recently, the cost of these biodegradable plastics, such as polyhydroxyalkanoates (PHAs), polymers made entirely by bacterial fermentation, has been one of the major limiting factors in their use. Producing PHAs in plants would significantly reduce the expense of manufacture compared with fermentation and make these biopolymers competitive with petroleum-based plastics for low-cost uses. Recently, efforts to produce PHAs in plants have been successful.

Bacterial genes that are necessary to synthesize polyhydroxybutyrate (PHB), a type of PHA, were transferred to *Arabidopsis thaliana* plants. The genetically engineered plants accumulated up to 14% dry weight of PHB without deleterious effects on plant growth or fertility and with a level of polymer yield that is considered commercially practical. Several companies are now pursuing development of PHA-producing genetically engineered oilseed rape (canola) crops.

### Environmental Risks

In initial prerelease testing, the interactions of such new genetically engineered organisms with nontarget organisms in the soil communities and contaminated landscapes must be carefully monitored to avoid potentially deleterious side effect. Also, it will be important to monitor the safe environmental disposal of plants that accumulate toxic materials.

## Novel Food Products

The objective of biotechnology and genetic engineering is to improve the food supply by increasing crop and livestock productivity, enhancing nutrient composition and availability, and improving food characteristics such as size, taste, and texture. The following are examples of the type of applications that have been achieved or are under way.

**Increased Crop and Livestock Productivity**

In the past, scientific breeders of plant and animals have utilized the rich genetic resources of cultivars and land races in crop improvement programs. It has been estimated that at least half of the increase in agricultural productivity realized this century may be directly attributable to "artificial selection, recombination, and intraspecific gene transfer procedures". Use of biotechnology and genetic engineering may result in even greater productivity gains through high-yield crop varieties, improved pest and disease control, the production of herbicide-resistance in crops, enhanced nutrient availability, and tolerance to a variety of environmental stressors. Biotechnology is also being used as a way to increase livestock productivity by enhancing milk and meat production. This is achieved by isolating the gene for the protein hormone (e.g., somatotropin) made in the anterior pituitary gland of animals and inserting it into bacterial cells, which results in an increased production of the hormone.

Applications of this technology include the injection of recombinant bovine somatotropin (rBST) into dairy cows to increase milk production and injection of recombinant porcine somatotropin (rPST) into growing pigs to increase carcass leanness and quantities of feed consumed per unit of output may benefit the environment by decreasing environmental pollution through reducing the quantity of feed provided; reducing the quantity of fertilizer and other inputs associated with growing, processing, and storing animal feed; and reducing animal waste products. Biotechnology may also play a role in improving crop plants, that in, turn, enhance secondary productivity in livestock. Researchers at the Commonwealth Scientific and Industrial Research Organization's division of plant industry in Canberra have found a way to insert a sunflower gene into the subterranean clover, a major constituent of Australian sheep pastures.

The new clover, with genes from sunflower that code for a protein in sulfur amino acids, provides sheep with a sulfur rich diet that is necessary for wool production. Creating such a transgenic clover resulted in about a hundred fold increase in the sulfur-rich protein. However, it has been estimated that (i.e. an increase in wool production by 5-10%), the level of sulfur-rich protein needs to be boosted another tenfold to have a substantial impact on wool production. The group is testing gene promoters that could achieve this. Producing the same amount of wool from fewer sheep could reduce soil erosion and other environmental pollution produced by these animals. Development in biotechnology and genetic engineering may also be able to improve ruminant nutrition by modifying the microbes that are involved in ruminal fermentation. The objective will be to find suitable foreign bacterial genes that can be inserted into ruminal bacterial organisms.

Techniques of genetic engineering are also playing an important role in increasing animal productivity by improving and developing vaccines and pharmaceuticals (e.g., fertility hormones). Hybridoma technology, which results in the generation of monoclonal antibodies by cell fusion procedures, will be increasingly useful in diagnosing specific diseases as well as in disease prevention and treatment. The broad range of potential vaccines for control of various diseases is especially promising because of their low environmental risks and excellent socioeconomic benefits.

## Enhanced Nutrient Composition and Availability

### *Enhanced nutrient composition*

Another application of biotechnology in agriculture is intended to improve the food supply by enhancing the nutritional composition of foods. In 1992, Monsanto was able to produce a genetically engineered potato successfully with an increased starch content. A higher starch content reduces oil absorbtion during frying and thereby lowers the cost of frying certain food products (e.g., French fries, potato chips) and reduces oil content in the finished product. Genetically engineered strains of oilseed rape (canola) are undergoing development at Calgene, where applications include high-stearic-oil-content margarine and edible canola oil with reduced saturated fat content. Through biotechnology, scientists, also hope to create foods that protect against cancer, heart disease, osteoporosis, and other life-threatening illnesses. For example, the National Cancer Institute (NCI) is currently working on increasing the amount of phytochemicals, which are linked to cancer prevention, in foods such as garlic, parsley, and citrus fruits. Government, university, and industry scientists are also developing cereal grains with increased amounts of both soluble and insoluble fiber to lower cholesterol and fight digestive cancers, respectively; milk with improved calcium bioavailability to help protect against osteoporosis; and vegetables with boosted levels of antioxidants (e.g., super carrots with five times the amount of β-carotene) to help reduce the risk of cancer.

### *Enhanced nutrient availability*

Biotechnology and genetic engineering can also be employed to enhance nutrient availability in agricultural systems. Genetic engineering has been applied to the problem of nitrogen fixation with a specific focus on the genetic makeup of organisms that fix nitrogen from the atmosphere and the genetic basis for the relationship between leguminous species (e.g., peas, beans, alfalfa, clover, peanuts) and the nitrogen-fixing bacteria that occupy their nodules. Scientists in China have recently developed recombinant strains of nitrogen-fixing bacteria that have a higher nitrogen fixation efficiency than traditional bacteria. These recombinant strains of bacteria have been spread over more than a million hectares of rice and soybean fields, and preliminary results show crop yield increases of 5-10%. In addition, a mechanisms of biological nitrogen-fixation similar to that of natural legumes eventually may be genetically engineered into plants such as wheat and corn. If this is achieved, the need for commercial nitrate fertilizers could be significantly reduced. However, because the molecular mechanisms required for symbiotic nitrogen fixation are complex (involving at least 17 genes), achieving this will require an investment in research over several decades.

### Other Improved Food Characteristics

Other applications of biotechnology to improve the food supply include the improvement of certain food characteristics such as size, ripeness, acidity or sweetness, taste, and texture. Genetic modifications have enabled the production of fruit that may have better taste and an enhanced shelf life through delayed pectin degradation or altered responses to the plant hormone ethylene. Among the first such novel products on the market was Calgene's slow-ripening Flavr Savr® tomato, which was approved for sale by the Food and Drug Administration in May 1994. This tomato has been transfected with an "antisense" gene responsible for the enzyme polygalacturonase, which solubilizes pectin. Because pectin degradation increases fruit ripening and decreases shelf-life, preventing translation of the message (by having the antisense message present) for polygalacturonase should retard fruit ripening and increase pre purchase and post purchase shelf life. DNA Plant Technology (DNAP) corporation's Endless Summer® tomato, which like Calgene's Flavr Savr has an antibiotic-resistance marker gene encoded into it, also incorporates a gene-splicing technique to retard ripening.

## Risks

### *Environmental risks*

The production of plants and animals to suit specific environments could lead to the transformation of more land presently occupied by natural ecosystems into agricultural land. The ecosystems of the tropics on the land not suited for agriculture of any kind are particularly vulnerable. Thus, the removal of yet more natural ecosystems would further deplete biodiversity. Genetic engineering may also favor monocultures that will threaten the global centres of ordered biodiversity.

### *Health risks*

One potential health risk is related to the use of genetically engineered animal hormones. One can illustrate using the case of rBST. The Food and Drug Administration has ruled that the presence of rBST in milk is safe for children and adults (FDA 1995). Even so, there have been question regarding the impact of this technology on animal and human health. Use of rBST in dairy cows increases the chances of bacterial infections and mastitis and also reduces the reproductive cycle in treated dairy cows. Milestone et al. (1994) reported that increased infections in cattle will require treatment with antibiotics. Although not all antibiotics appear in milk, some do. Thus, if more antibiotics are used, an indirect risk to humans may arise because some residues may remain in the milk. A second potential human health concern is the risk of introducing allergens into the food supply.

A recent study by Nordlee and colleagues demonstrated the transfer of a major food allergen from Brazil nuts to transgenic soybeans during the development of a genetically engineered crop variety. Furthermore, although only a dozen foods may produce allergic reactions—mainly protein foods—biotechnology allows for nontraditional food proteins (e.g., moths, insects) to be present for which no knowledge currently exists regarding their allergenic or nonallergenic qualities. A third potential human risk relates to the increasing prevalence of antibiotic-resistant and disease-causing bacteria. Because antibiotic-resistant genes are the most commonly used type of selectable marker in genetic engineering and are rarely deleted from the resulting organisms, incorporation of antibiotic-resistant markers into the genetic material of human pathogens could pose risks by increasing the prevalence of antibiotic-resistant and disease-causing bacteria.

In 1989, dozens of people died and thousands others were crippled after consuming a batch of synthetic L-tryptophan produced using genetically-engineered bacteria. Although the exact cause of illness will never be known, it has been reported that a specific impurity stemming from the bacterial strain may be cause of the syndrome. A fourth health concern is that genetically engineered crops may be able to transfer their foreign gene(s) to other unrelated microorganisms. Genetically engineered oilseed rape (canola), black mustard, thorn-apple, and sweet peas all contain an antibiotic-resistance gene and were grown together with fungus *Aspergillus niger*, or their leaves were added to the soil.

The fungus was shown to have incorporated the antibiotic-resistance gene in all coculture experiments. Other health risks created through the use of biotechnology and genetic engineering include the ability of genetically engineered organisms to survive and harm nontarget organisms, the creation of new toxic organisms, or both. The risk of a new host being infected by a virus or recombining to form a more deadly, virulent virus must be investigated further. For example, the risk of recombination between the engineered vaccine virus and other orthopox viruses endemic in wildlife, such as the cowpox virus, still needs to be investigated accurately.

### *Social and economic impact*

Proponents often cite biotechnology and genetic engineering as the way to help solve the world's food problems (i.e., improve agriculture in developing countries). However, some critics who are skeptical about of ability of biotechnology and genetic engineering to increase food production project there will be about a 1% increase in crop yield on the basis of the anticipated contributions of biotechnology during the next two decades. In addition, no increase in crop yields is projected for Africa because no major advances in biotechnology are expected to be applied in Africa in the near future. Some persons further question why, after 20 years of research, biotechnologists have not produced a single high-yielding variety of wheat, rice, or corn. The answer, according to plant scientists, is that plant breeders using traditional techniques may have largely exploited the genetic potential for increasing the share of photosynthate that goes into the seed. Others feel that the present channeling of funds to expensive biotechnology projects diverts scarce resources from other research that could focus on more practical solutions to pressing social problems such as hunger and food insecurity.

Some authors have also observed that many of the crops being engineered for herbicide resistance belong to the group of key crops in Western agriculture. This circumstance may reflect the domination of the majority of the biotechnology industry by transnational companies (TNC) in the developed world whose business it is to generate profits. If sustaining the Third World is the largest of genetically engineered crops, other vegetables and crops have to be considered. Also, helping these countries bypass expensive, high-input crop production and move their traditional agriculture toward low-input sustainable practices is desirable as well.

## Transgenic Animals

### Benefits

Among the benefits of producing genetically engineered or transgenic animals are the ability to manufacture more cost-effective drugs (including vaccines), new animals models to study human disease, human tissue and organ harvesting, and improvements in the food supply (e.g., modifying the shape, size, or nutritional quality of animals). In the early 1980s, the first successful experiments creating transgenic vertebrates were reported. Palmiter et al. (1982) created a transgenic mouse by transferring a growth hormone into the embryo of the mouse. In 1988, a patent was awarded for a mouse that was genetically engineered with human genes designed to be tumorigenic. This mouse, commonly referred to in the scientific literature as the "oncomouse" (oncology is the study of tumors), prompted the first patent issued for a transgenic animal.

Transgenesis, or the transfer of genes across species lines, opened the way for animals to be used as an alternative to tissue-culture productions of human protein. In 1987, a transgenic mouse was created that demonstrated the viability of tissue-specific expression of foreign proteins. As reported by Krimsky and Wrubel (1996), "The mouse was genetically engineered to make the clot-dissolving factor tissue plasminogen activator (TPA), which is viewed as a highly promising drug for the treatment of coronary heart disease and a strong competitor of the widely acclaimed streptokinase." Krimsky and Wrubel (1996) also note, "Finnish researchers have developed a genetically-modified cow that purportedly can produce milk containing large amounts of erythropoietin (red cell growth factor) used to treat anemia. If successful, this method will replace the costlier cell culture techniques. Other more remote applications of transgenic animals include human blood and organ production."

Transgenic animals under development include swine with the human growth hormone gene, genetically engineered livestock designed to tolerate extreme climatic conditions, transgenic sheep that grow faster than normal sheep, engineered sheep that secrete insect repellent and produce moth-proof wool, and genetically engineered sheep and cows that product milk consumable by lactose-intolerant individuals. Many different species of fish are being genetically engineered to increase fish size and growth rate or improve survival in new environments. Genetic engineers have turned much of the attention toward finfish and shellfish. Many species of transgenic fish have been grown in the laboratory.

Fast-growing Pacific salmon have been engineered by various groups of researchers worldwide. Scientists attached a switch to a growth hormone gene from coho salmon and injected the transgene into chinook salmon eggs. On average, the transgenic salmon grew to be eleven-fold heavier than their age-mates. Fast-growing fish have also been produced outside the laboratory. In 1991, transgenic carp were tested in a high security pond at Auburn University. These fish were fitted with a growth hormone gene from rainbow trout that enabled them to grow 40% faster than normal. Researchers have also identified a protein in winter flounder, which has recently been transferred into Atlantic salmon, that prevents fish blood from freezing. Transgenic salmon with this trait could potentially be raised in sea pens farther north, where the species could not otherwise live.

## Risks

### *Environmental*

Very real environmental risks may be associated with transgenic animals—particularly with certain applications, such as transgenic fish. For example, transgenic fish have the potential to disturb ecosystems seriously by gaining a competitive advantage in the wild ecosystem. A fast-growing transgenic fish could assume a higher than usual position in the food chain because of its greater size and ability to compete for food, which could harm native species, or a freeze-tolerant transgenic fish raised in the north could escape into a geographic area from which it was previously excluded, and cause competition that would harm native species. Fertile transgenic fish could also successfully invade ecosystems, thus exacerabating the present problem with exotic invaders in aquatic ecosystems.

### *Health*

Using transgenic animals to harvest blood, tissues, or organs may create certain health risks. For example, take the case of deriving human hemoglobin from transgenic pigs. Human hemoglobin must be separated from that of animal to ensure the purity of the product. Also, how humans will respond to such animal-derived human hemoglobin is still untested.

### *Social and ethical*

Finally various social and ethical risks are associated with the production and use of transgenic animals. Certain animal rights groups (e.g., The Human Society) question whether it is morally or ethically correct to turn animals such as mice, pigs, and sheep—which unlike plants and bacteria are sentient begins—into biomachines for the manufacture of proteins and other biological materials. Some groups may also find it unethical to introduce human genes into livestock and plants. This application of genetic engineering could raise ethical concerns and seriously undermine the public's perceptions of biotechnology. Religious groups may also have certain conflicts as to whether this new technology is compatible with each of their traditional norms. There may also be a conflict of interest within the value and belief systems of those groups for whom it is abhorrent to manipulate, control, experiment with, or consume animals of any type.

## General Risks of Releasing Genetically Engineered Organisms into the Environment

### Single-gene Changes and Pathogenicity

Most single-gene changes are probably not likely to affect the pathogenicity and virulence of an organism in nature adversely. However, some gene changes may have detrimental consequences.

Certain genetic alterations in animal and plant pathogens, for example, have led to enhanced virulence and increased resistance to pesticides and antibiotics. For instance, some oat rust microbes, initially nonpest genotypes for a particular oat variety, became serious pest genotypes after a single gene change allowed the rust to overcome resistance in the oat genotype. An important fungal disease of rice, rice blast, has been demonstrated to have genotypes with single-gene changes that cause the fungal organism to be potentially pathogenic to rice cultivars.

A similar phenomenon of single-gene changes resulting in pathogenicity has been documented with a related fungal pathogen that infects weeping love grass. This phenomenon has led plant pathologists to develop the "gene-for-gene" principle of parasite–host relationships in which a single mutation in a parasite overcomes single-gene resistance in the host. Furthermore, numerous instances have been documented in which insects, through a single-gene change, have overcome resistance in plant hosts or have evolved resistance to insecticides. More than 500 species of arthropods have developed resistance to pesticides.

## Threats from Modified Native Species

Lindow (1983) has reported that there is little or no danger from the ice-minus strain of *Pseudomonas syringae* (Ps) because Ps is a native U.S. Species that produces related phenotypes in nature. Other investigator have demonstrated that there are different genotypes of Ps, and some of these genotypes have genes for pathogenicity. Because some native species have the ability to alter their interactions within an ecosystem, the genetic modification and release of native species into the natural ecosystem may not always be safe. For example, from 600 to 80% of the major insect pests of U.S. and European crops, respectively, were once harmless native species in the United States and Europe. Many of the insects moved from benign feeding on natural vegetation to destructive feeding on introduced crops. For instance, the Colorado potato beetle moved from feeding on wild sandbar to feeding on the potato that was introduced form Peru and Bolivia. This insect has become a serious pest of the potato in the United States and Europe.

## Intentional Introduction of Crops Plants and Animals

Some proponents of biotechnology suggest that the intentional introduction of foreign plants and animals into the United States is a good model for predicting potential problems arising from biotechnology. If so, there is reason for concern because several serious problems have resulted from the intentional introduction of what were believed to have been beneficial crops and animals. Genetic similarities between many of the crops and weeds are evident from the fact that 11 of the 18 most serious weeds of the world are crops in other regions of the world.

Of the several thousand crops that were intentionally introduced into the United States, 128 species of agricultural and ornamental plants species, like Johnson grass, are among the most serious weed species in the United States—especially in the southeastern United States. Johnson grass was introduced as a forage for livestock before it escaped and became a weed pest. This pattern of native species and introduced plants species in the United States is not unique. For example, Florida has only 2525 indigenous plant species, but approximately 25,000 plant species have been introduced and are under cultivation there. An additional 925 species of exotic plants are established in nature in Florida, and several of these are currently displacing native plant species.

One exotic pest that is displacing native plant species is the melaleuca (*Melaleuca quinquenervia*) tree introduced from Australia. This clearly illustrates the threat to our natural ecosystem when so-called beneficial organisms are introduced and released into nature. The picture in Florida for insects is quite different than for plants. The great majority of insects species (11,512) are native, with nearly 1000 immigrant species. A relatively small number (42) of insect species are established in nature. One of these species that has established itself and is a serious pest in nature is the imported fire ant. Furthermore, 9 out of 20 introduced domestic animal species in the United States have displaced or destroyed native species.

These introduced domestic animals, including donkeys, horses, and goats, have become serious ecological pests. A total of 10 other introduced animals, including mammals (e.g., mongoose and wild boars) and birds (e.g., English sparrow and mynah), have become pests. Furthermore, at least 70 species of fish have been introduced and have become established in the U.S. aquatic ecosystems. These 70 species represent about 10% of all U.S. fish species. A total of 5 introduced fish species have become pests, displacing and reducing the number of native species and, in other cases, altering the habitat and making it uninhabitable for fish and other species. In addition to these intentionally introduced fish, there is concern about the introduction of transgenic fish and the potential ecological effects of these engineered fish in aquatic ecosystem in the Untied States. This overall history suggests that the introduction of many types of foreign organisms in the ecosystem may have major negative impacts on many of the 500,000 beneficial plants and animals in the United States.

### Are Ecological Niches Filled?

An estimated 1500 exotic insect species have been introduced and are established in the United States, and a few of these (17%) have become pests or have a negative impact on native species. This observation indicates that few of the niches in natural ecosystems are filled. There is ample opportunity, therefore, for may species to become established in the United States. Thus, although the argument that engineered organism swill not become established owing to competition with native species may apply in a few cases, most often this is not a valid argument.

## Uses of Biotechnology and Genetic Engineering for a Sustainable Agricultural System

Acceptable and potentially sustainable options for agriculture that could be derived from biotechnology and genetic engineering have been put forward by several authors. Some of the desirable areas of development for such technologies that have the potential to benefit agricultural sustainability, the integrity of the natural environment, and the health and safety of society are discussed in the following paragraphs.

### Enhancing Crop Resistance to Pests

Approximately 500,000 kg of pesticides are applied each year in U.S. agriculture, and many nontarget species beneficial to the environment are negatively affected. Genetic engineering targeted for pest control could diminish the need for pesticides. Resistance factors and toxins that exist in nature can be used for insect pest and plant pathogen control. For example, more than 2000 plant species are known to possess some insecticidal activity, and approximately 700 natural substances in bacteria, fungi, and actinomycetes have fungicidal activity.

Traits for resistance to different insect pests and diseases already exist in many cultured crops, including corn, wheat, barley, soybeans, beans, apples, grapes, pears, tobacco, tomatoes, and potatoes. Although some resistance characteristics have been reduced or eliminated in commercial crops, they still can be found in related wild varieties, which provide an enormous gene pool for the development of host-plant resistance. For example a wild relative of tobacco that produces a single acetylated derivative of nicotine is reported to be 1000 times more toxic to the tobacco hornworm than cultivated tobacco. Transferring this toxic gene to nonfood crops, such as ornamental shrubs and trees, may protect them from certain insect pests. In addition, thionins, proteases, lectins, and chitin-binding proteins that are often present in plants, especially in the seeds, help control some pathogens and pest insects in wild plants. For example, it has been shown that a cowpea protease inhibitor found in the cowpea *Vigna* spp. can now be engineered into tobacco. Laboratory trials also indicate that the cowpea protease inhibitor provides protection against the cotton budworm (*Heliothis virescens*), which is a major pest of tobacco, cotton, and maize.

**Development of Perennial Crops**

At present, the major cereal crops of the world are annuals. The conversion of annual grains to perennial grains by genetic engineering will reduce tillage and erosion and conserve water and nutrients. Such crops will decrease labour costs, improve labour allocation, and, overall, improve the sustainability of agriculture. Energy efficiency in cultivation of perennial cereal crops will be greatly superior to that of annual crops.

**Improved Botanical Pesticides**

Only limited quantities of botanical pesticides, such as pyrethrums, are now used in developed countries in place of some synthetic pesticides. However, in some developing countries, including China and India, botanical pesticides such as neem are effectively used. Increasing the effectiveness of need and other available botanical pesticides by genetic engineering would be an asset to farmers because these substances are relatively effective and safe.

**Bioindication Needs for Sustainable Use of Genetically Engineered Plants**

Bioindication is a strategy that adopts and assesses biological units, species, assemblages of species, and ecosystem models to determine the impact of a selected contaminant such as pesticide residues or fertilizers on the environment. This strategy is aimed at using biological nontarget organisms, both in microcosm-modeled and in field arenas, to assess the environmental problems created by adopting certain new management techniques in agroecosystems, including genetically engineered organisms. As observed by several authors, little work has been dedicated to assessing the true environmental impacts of genetically engineered crop plants. For example, there are no data on pollinators of engineered plants modified with the BT δ-endotoxin. It would be rather useful to work with the nontarget species linked to the rural landscapes in which the genetically modified crops are expected to be introduced.

## Public Perceptions of Biotechnology

Current public opinion polls show that consumers find some applications of biotechnology and genetic engineering more acceptable than others. For example, in a national random telephone survey conducted in the United States by Hoban and Kendall (1992), 66% of respondents considered plant-to-plant transfers acceptable, 25% found animal-to-plant gene exchanges acceptable, 40% found animal-to-animal gene transfers acceptable, and only 10% approved of human-to-animal gene transfers.

A survey of public attitudes toward biotechnology in the United Kingdom revealed that a large percentage of respondents would accept genetically manipulated foodstuffs under the condition they were confident about testing and that the food product(s) look and taste the same or better. One-half of respondents felt it would be a good thing to use genetic manipulations to solve the food problems in the Third World, whereas relatively few people felt it would be a good thing to use genetic manipulation to provide or improve the food supply in the Western world.

A majority of respondents (79%) also felt that it would be a bad thing to use genetic manipulation for products they did not feel were needed. Consumer surveys in the United States also suggest that the public is skeptical and cautious with regard to certain applications of biotechnology in agriculture and food production. Consumer apprehensions center around biotechnology's perceived unpredictability, risks to the environment, and moral and social questions. In Europe, public attitudes toward genetically engineered foods appear less favourable than in the United States. Recently, the European Commission has approved mandatory labeling of all genetically modified organisms (GMOs). The European community states that the intent of the label is to serve as a source of information for consumers, not as a warning. Results from consumer surveys in the United States, Canada, and the United Kingdom also indicate the most consumers are in favour of labeling genetically engineered foods.

## USE OF BIOTECHNOLOGY AS A WAY TO PRESERVE BIODIVERSITY

### Benefits

Proponents of biotechnology claim that use of biotechnology and its subdiscipline of genetic engineering is needed to renew the momentum of plant and animal genetics and is one of the factors that has contributed to food production gains achieved thus far. Advocates also claim that biotechnology has the potential to enhance species preservation, increase the value of biodiversity, and promote ecosystem conservation through decreased environmental degradation.

#### *Enhance species preservation*

The International Board for Plant Genetics (IBPGR) as well as many others has taken up the job of collecting critically valuable germ plasm. The IBPGR gene banks now have more than a half a million plant "accessions" representing thousands of varieties and hundreds of different species. Biotechnology has encouraged the collection of genes through germ plasm secondary to its being viewed as a valuable and prosperous activity. Utilizing biotechnology for collecting germ plasm may help ensure the world's biodiversity is maintained by fostering preservation of species that are in danger of becoming displaced by new varieties.

#### *Increase the value of biodiversity*

Biotechnology may increase the value that is placed on biodiversity through increased returns on investments in research and development of biotechnologies that may generate animal and crop breeds of potential value. The present economic benefits of biotechnology products are significant and are conservatively estimated to be between $2-3 billion/yr. Nearly half of the current economic benefits relate to agriculture with significant benefits to the pharmaceutical industry. For example, biotechnology, in the form of bioassays, has reduced the time and cost of screening for pharmaceutical and other uses and has thus increased the value of the underlying genetic resources. For this reason, pharmaceutical companies are becoming more interested in the potential biochemical properties of tropical species and varieties for developing new drugs. On the basis of data from Costa Rice, Aylward (1993) estimates that the net private returns to pharmaceutical companies prospecting for biological resources are around $4.8 million per new drug per year. Over 50% of these royalty returns could realistically be allocated to biodiversity protection in Costa Rice.

### *Conserve ecosystems through decreased environmental degradation*

The improvement of crop and livestock productivity could, in principle, create an environmental advantage. Less land, particularly less marginal land, would need to be cultivated, thereby reducing problems such as soil erosion and desertification and promoting ecosystem conservation. However, this would require more substantial productivity gains than have been achieved thus far with crop cultivars that have been produced though biotechnology and genetic engineering. In addition to high-yield agriculture, biotechnology may produce high-yield forestry, which could not only speed up growth of tree crops such as rubber, pulpwood, and cocoa but would also help preserve rapidly depleting forests. Although not yet achieved, biotechnology could also help promote environmental management by reducing the use of artificial fertilizers and chemicals and fossil-fuel consumption, which leads to environmental and biodiversity destruction.

## Risks

On the other hand, biotechnology and genetic engineering may produce various environmental and social and ethical risks that could ultimately lead to further habitat destruction and depletion of genetic resources. These potential risks include further depletion of biodiversity, harm to nontarget organisms, and exploitation of farmers in developing countries by translation companies.

### *Environment*

#### *Further depletion of biodiversity*

Use of biotechnology could actually lead to the transformation of more land and thus cause the removal of yet more natural ecosystems into agricultural land, which would further deplete biodiversity. For example, the ecosystems of the tropics, where land may not be suitable for agriculture of any kind, may be particularly vulnerable. Encouraging use of certain genetic resources in the production of pharmaceuticals may also lead to a further depletion of biodiversity. Biotechnology and genetic engineering may also favor monocultures and threaten the global centers of ordered biodiversity—the basis for ecological stability, which has already been seriously undermined primarily as a result of global industrialization, urbanization, and overexploitative, agricultural practices.

*Harm to nontarget organisms*

Other environmental risks created through the use of biotechnology and genetic engineering include the ability of genetically engineered organisms to survive and harm nontarget organisms and the creation of new toxic organisms. The genetic engineering of viruses and bacteria could lead to the accidental production of toxic or environmentally harmful strains. Not all ecological experiments are successful with their potential predictions, and such errors of judgment have had negative effects on the environment in the past.

### *Socioeconomic*

*Transnational companies (TNCs) may exploit farmers in developing countries*

One social concern of biotechnology is that TNCs are building monopolies over transgenic seed production and may eventually disadvantage poor farmers in developing countries. For example, the packaging of seeds with engineered herbicide resistance and the herbicides by agrochemical companies could be viewed in this light. The situation is made even more complex because the majority of the genetic resources, and thus biodiversity, on which biotechnology depends are found in developing countries. A second concerns is that because genes extracted from ecosystems in developing nations can be engineered and turned into valuable assets it will be possible for the genes to be patented by developed nations, which will result in developing-world farmers paying for products that original from their nation's own resources.

### Risk Assessment and Policy Recommendations

If biotechnology is to contribute to sustainable agricultural development, polices must be adopted to ensure that the profit generated by biotechnological research and development is invested in the conservation of habitats that produced it and that prospecting ventures contribute economic benefits that build technological capacity in the country of origin. An example that illustrates how this can be done is the partnership between Costa Rica's National Biodiversity Institute (INBio) and Merck and Company, Ltd., by which Merck provide INBIo with a $1.1 million dollar budget and INBio provides Merck with plant and animal extracts from biologically diverse, undeveloped areas.

In addition, INBio also receives a share of the royalties on any products that are ultimately developed. Mexico, Indonesia, and Kenya are establishing similar bilateral agreements. A second way that biotechnology can contribute to sustainable agricultural development is through adoption of an International Biosafety Protocol commissioned by article 19 of the Convention of Biological Diversity (CBD), which was introduced and opened for signature at the Earth Summit meeting in Rio de Janerio in 1992. Because genetically modified organisms (GMOs) are new and many pose novel risks, their future impact on the environment and health is uncertain.

Many scientists and organizations, including the authors of this chapter, believe the international community would benefit from a legally binding protocol that sets basic standards for the release and export of GMOs and would prevent further damage to biodiversity and the Earth's ecosystems. Certain nongovernmental organizations (NGOs) have already begun to draft such a protocol. The drafted protocol would require exporters of biotechnology products to submit complete safety information on a case-by-case basis, establish an independent international body of experts to conduct risk assessments and make decisions on all transboundary trade of genetically modified organisms (GMOs), include public participation at every step of decision making, require mandatory labeling for genetically engineered food products, provide technical and analytical support for member countries, and establish liability standards.

The protocol also calls for risk assessment to include social and cultural studies and states that genetic diversity "is dependent on the socioeconomic conditions of the peoples maintaining it". However, the United States would most likely only adopt a significantly modified position, that is one that is science-based and within a framework of risk assessment and management that has been proven adequate in the United States, Europe, and elsewhere. Certain other in the scientific community do not support the adoption of the CBD and its legally binding International Biosafety Protocol and believe that regulation of biotechnology would serve as a hazard to the diffusion of biotechnology in the developing world; stifle the development of certain applications of biotechnology such as those that can assist in toxic waste removal, water purification, and displacement of agricultural chemicals; and would not likely meet the protocol's goal of safety enhancement cost-effectively.

## Conclusion

Techniques of biotechnology and genetic engineering, if adopted responsibly, have the potential to increase productivity in crops and livestock, control pests, produce new food and fiber crops, and develop effective medicines. Potential environmental and economic benefits from biotechnology include the reduction of fossil fuel in agriculture and forestry through improved nutrient availability in crops and livestock, use of fewer artificial inputs (e.g., synthetic nitrogen fertilizers, insecticides, and fungicides), and more cost-effective and environmentally friendly waste management practices such as bioremediation. If realized, these improvements will help protect ecological systems by reducing habitat degradation.

Although biotechnology and genetic engineering can be expected to provide major benefits to agriculture and the environment, risks with the use of this technology should also be recognized.

Environmental risks include: the potential to alter basic interactions in natural ecosystems; create plants that become new weeds; release pest control organisms that evolve resistance or harm nontarget organisms, or both; the deplete biodiversity further.

Potential health risks include: the possibility to introduce new allergens into the food supply, increase levels of antibiotic residues in the food supply, increase the prevalence of antibiotic-resistant bacteria, and create new virulent strains of bacteria and viruses.

Socioeconomic risks include: the ability of transnational companies (TNCs) to create monopolies and exploit farmers in developing countries. Ethical risks include issues related to the inhumane treatment of animals and conflicts arising in value and belief systems of certain individuals and organizations.

The public appears to find certain applications of biotechnology and genetic engineering acceptable, including plant-to-plant gene transfers and use of this technology to solve food problems in developing countries. However, other applications of biotechnology, such as animal-to-plant gene exchanges are reported to be less acceptable. The majority of consumers across international borders support mandatory labeling of genetically engineered foods. Because the goal of biotechnology is to reduce rather than increase risk in the food supply, a more effective international regulatory policy is needed. Adoption of an International Biosafety Protocol, as commissioned by article 19 of the Convention on Biological Diversity (CBD), would help reduce the new and novel risks associated with the use of biotechnology and genetic engineering.

# Chapter 4

# Role of Cytokinin

For most of medical history, therapeutic intervention has relied on exogenous factors. These treatments have run the gamut from magical incantations, to animal and plant derived substances, and finally to rational drug design by combinatorial chemistry. Success in treatment has therefore depended on placebo effect, fortuitous similarities in ligands and receptors (e.g., poppy-derived opiates and the opioid receptors in the brain), or a grasp of biological mechanisms at the molecular level. Paradoxically, this increased understanding of the biology of health and disease, as well as the tools that were developed to gain this understanding, has led to the possibility of utilizing the body's own endogenous factors to treat disease and restore health. One potential target of this therapy is the group of factors collectively known as cytokines. In this chapter we will describe cytokines and their diverse use in human therapeutics, and provide an overview of the multitude of experimental and clinical approaches that have been designed to tap into the body's own healing power. Along the way we will also describe some of the known and potential disadvantages of this existing new therapy. This review is not meant to be a comprehensive listing of the results of individuals studies but rather a broad survey of the types of approaches used.

## Basics of Cytokine Biology

The majority of cytokines (also known by their previous name, lymphokines) are glycoproteins secreted by cells, although membrane-associated forms have also been described. Cytokines are generally not produced constitutively but rather in response to cellular activation. Cytokine production is highly regulated in both paracrine and autocrine fashions at the transcriptional and posttranscriptional levels. In addition, the short half-life of cytokine mRNA suggests selective degradation of the message as a result, evidently, of a particular sequence common to many cytokines and proto-oncogenes.

Finally, a variety of mechanisms appear to ensure that cytokines remain compartmentalized. These mechanisms all work together to limit the biological activity of cytokines. Perhaps the most important biological feature of cytokines is that they are highly pleiotropic and redundant in function. That is to say, most of the cytokines described to date have multiple actions, and these actions often overlap with the actions of other cytokines. In addition, cytokines frequently work via cascading mechanisms that allow them to interact with each other synergistically and antagonistically. These features are critical from a therapeutic consideration because the ultimate consequences of manipulating any single cytokine must be evaluated in the context of this overall network of factors.

Other features of cytokines important form a clinical standpoint are that, in most cases, they act primarily at a local level and that they are rapidly cleared from the circulation. A notable exception to this level and that they are rapidly cleared from the circulation. A notable exception to this rule is interleukin-6 (IL-6), whose importance in regulation of the acute phase response has led to a variety of "chaperone" mechanisms for regulating its systemic bioavailability. Cytokines interact with specific receptors grouped into the hemopoietin superfamily the tumor necrosis factor (TNF) family, the immunoglobulin superfamily, and the tyrosine kinase family. These receptors are generally membrane-bound molecules, although soluble receptors have also been described.

Many cytokine receptors share several similar characteristics, including a subunit structure and an association with signal-transducing elements within the cell. Moreover, several of the receptor subunits are shared between various cytokine receptors, which possibly contributes to cytokine pleiotropy and redundancy. A frequent conceptual mistake is the belief that cytokines are an exclusive product of the immune system. In fact, certain cytokines are phylogenetically ancient, genetically conserved, and highly pleiotropic molecules (probably the best examples are IL-1 and TNF); forms of these molecules are found in invertebrates lacking a true immune system.

Furthermore, IL-1, TNF, and related molecules are intimately involved in the process of apoptosis, a fundamental biological process. Put another way, cytokines should be recognized for their role as conveyers of bio-information rather than as simple effector molecules involved in a single process. This brings us to a crucial detail; cytokines should not, indeed cannot, be seen as operating in a biological vacuum. Rather, as suggested cytokines may be thought of as a member of a triad including (at a minimum) neuropeptides and hormones and possibly the so-called peptide growth factors as well. It is now well established that these molecules form a complex network responsible for regulating many physiological processes.

Certain cytokines (e.g., IL-6) mediate autocrine functions in the nervous and endocrine systems. Likewise, cells of the immune system produce hormones (e.g., prolactin) and neuropeptides (e.g., endorphins); in addition, receptors for hormones and neuropeptides are found on lymphocytes. Thus, given the redundant and pleiotropic nature of cytokines themselves as well as their role as regulatory molecules for various physiological processes, it becomes apparent that developing therapeutics based on modifying the functions of cytokines is far from straightforward. As detailed later in this chapter, toxicity and other unintended consequences are often the result of initial forays into this area. Nevertheless, great advances have been made and continuous apace. Table 4.1 illustrates a suggested categorization of the various human cytokines. Cytokine nomenclature is fraught with confusion; beginning with interleukin-1, many newly discovered cytokines were named sequentially regardless of their function or genetics relationship to other cytokines.

In most current classification schemes, cytokines are grouped according to function or genetic similarity to each other. Table 4.1 is not meant to be inclusive; on the contrary, many excellent classification schemes based on function have previously been published. A brief explanation of the classification scheme used here may be helpful. In general, the term *interleukins* refers to cytokines exerting their primary effect on cells of the immune system; IL-1 and IL-6 are notable exceptions to this broad categorization because they mediate a tremendous number of effects of besides immunomodulation. *Chemokines* are small polypeptides that are important not only in immunity and

**Table 4.1. The (abbreviated) universe of human cytokines**

| Interleukins (IL) | Chemokines | |
|---|---|---|
| IL-1α, β | α (C-X-C) | β (C-C) |
| IL-1RA | IL-8 | RANTES |
| IL-2 | gro (α, β, γ) | I-309 |
| IL-4—IL-7 | GCP-2 | eotaxin |
| IL-9 —IL-18 | ENA-78 | MIP-1 (α, β) |
| | SDF-1 | MCP (1, 2, 3) |
| **Colony-stimulating factor (CSF)** | **Hematopoietins** | |
| IL-3 (Multi-CSF) | Erythropoietin | |
| G-CSF | Thrombopoietin | |
| M-CSF | Stem cell factor | |
| GM-CSF | fit3 ligand | |
| **Tumor necrosis factors (TNF)** | **Interferons (IFN)** | |
| Tumor necrosis factor (TNF-α) | Type I (IFN-α, IFN-β) | |
| Lymphotoxin (TNF-β) | Type II (IFN-γ) | |
| **Miscellaneous** | | |
| Oncostatin-M | | |
| Leukemia inhibitory factor | | |
| Transforming growth factor beta | | |

inflammation but in other regulatory processes as well. *Colony stimulating factors* (CSFs) are important as mediators of hematopoiesis but also exert effects on the immune system. *Hematopoietins* also regulate the function of bone marrow but have fewer effects on the immune response.

*Tumor necrosis factors* (TNFs) regulate a panoply of biological effects. *Interferons* were originally named on the basis of their ability to interfere with viral replication, and thus they function as important mediators of host defense; however, they also affect various other processes. Finally, we may lump together *miscellaneous* cytokines that do not fit quite so neatly into the scheme owing to their multiplicity of action. Not included in this scheme are the various *peptide growth factors* such as epidermal growth factor and insulin-like growth factor—whose functions appear to be primarily related to growth, tissue repair, and homeostasis—but that may function as cytokines in certain circumstances. All of the these cytokines classes, to varying degrees, represent viable targets for therapeutic manipulation.

## Cytokines in Human Health and Disease

Given the ubiquitous role of cytokines in normal human biology, it is not natural that disruption in cytokine levels could be either the cause or effect of human disease. A major advance in understanding the role of cytokines in disease, and particularly in diseases incorporating an immune component in their etiology, was first described by Mosmann et al. (1986). These investigators found that T-helper lymphocytes could be categorized functionally (although not, as yet, phenotypically) into at least two subsets on the basis of their particular pattern of cytokine production.

Originally described in vitro using T-cells clones, these subsets were named T-helper-1 and T-helper-2 (subsequent to this, the nomenclature TH1/TH2, T1/T2, and Type 1/Type 2 has also been used: TH1/TH2 will be the preferred usage in this chapter). Building on the data collected from in vitro clones, researchers found that these patterns of cytokine production resulted in vivo in rodent models. More recently, the TH1/TH2 dichotomy has been confirmed in humans. In general, TH1 cells predominantly produce interferon-gamma (IFN-γ), IL-2, and TNF, whereas TH2 cells primarily make IL-4, IL-5, IL-10, and IL-13; these patterns appear to be essentially the same in humans. Interleukin-2 acts as an autocrine growth factor for TH1 and TH2 cells are much

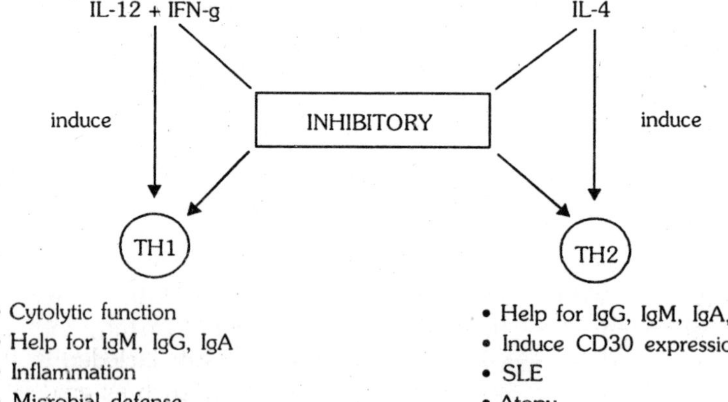


*Fig. 4.1.*

sensitive to this activity. Interferon-gamma acts to suppress the growth of TH2 cells. Conversely, IL-4 acts as the autocrine growth factor for TH2 cells, whereas IL-10 suppress the growth and function of TH1 cells.

Various other cytokines are produced in common by these subsets, albeit at relatively lower levels. It should be noted that these distinctions are somewhat fluid, for clones have been isolated in both human and laboratory animals cells that display intermediate cytokine production profiles. Moreover, CD8 (cytotoxic-suppressor) T cells also produce cytokines, as do macrophages and B cells, thus complicating the elucidation of a "typical" cytokine response. However, the general pattern described above for T-helper cells appears to hold true in most cases. The induction of a TH-1-type response versus a TH-2 type response (i.e., a state characterized by the preponderance of either TH1- or TH-2 type cytokine production) depends on a variety of factors, including the type of antigen-presenting cell, the type and amount of antigen, and factors in the cellular milieu such as other cytokines and hormones.

This particular pattern of stimulatory-inhibitory-autocrine functions is now recognized to form the basis for differential immune responses as well as certain pathologies. For example, in normal immune responses TH1 cytokines tend to favour cytolytic (cell-mediated) reactions and the activation of macrophages, whereas TH2 cytokines provide T-cell help for the production of various antibodies. On the other hand, TH1 responses are usually associated with autoimmune diseases (e.g., multiple sclerosis and rheumatoid arthritis), whereas TH2 responses include atopic conditions, reduced response to infectious challenge, and even successful pregnancy. In current and future therapies involving cytokines, it will be of vital importance to understand this interlocking network of cytokines in greater detail to allow possibly for the direct manipulation of immune responses.

**Immunomodulation**

Ironically, although cytokines are usually associated with the immune system, direct modulation of the immune response has not been the principal therapeutic application of these factors to data. A wide range of options are theoretically available for modulating the immune system, and most of these options are described in this chapter.

### *Neoplastic disease*

One of the most thoroughly investigated therapeutic uses of cytokines thus far has been in the treatment of neoplastic disease. Owing to the intrinsically complex nature of neoplastic disease, it follows that the use of cytokines for treatment is equally complex. Therapeutic options to data include disruption of autocrine growth, enhancement of natural immune resistance, the use of cytokines in combination with other drugs, and direct tumor cytotoxicity by cytokines. It is recognized that many tumors, particularly neoplasias of the hematopoietic system (e.g., leukemia) may involve the autocrine growth of cytokine-responsive or cytokine-producing cells. For example, IL-6 is thought to have the potential to induce polyclonal B-cell activation. In such cases, selective reduction in the levels of certain growth factors could theoretically be beneficial (although clinical trials employing extracorporeal removal of cytokine have yet to demonstrate clinical improvements).

Conversely, deficiencies in cytokine production may impair the function of cellular immune components required for resistance to tumors, such as cytotoxic T lymphocytes or natural killer cells. In such cases, the administration of carefully selected cytokines or cytokine combinations may boost a failing immune response. An intriguing concept that has been explored is the combined administration of cytokines with cytotoxic drugs. Experimental data suggest that administration of cytokines may result in enhanced susceptibility to cytotoxic drugs under certain circumstances. Certain cytokines may also exert a direct cytotoxic effect on tumor cells—either as single agents or in combination. Finally, a relative successful use of cytokines in oncology has been the repopulation of bone marrow following transplant subsequent to chemotherapy or radiation.

### *Transplantation*

Another successful therapeutic use of cytokines has been as an invaluable adjunct to transplantation. Ironically, successful treatment may take the form of either cytokine suppression or cytokine addition. In the former situation, cytokines are highly active participants in the rejection process responsible for the failure of organ transplantation and may contribute to engraftment failure by a variety of mechanisms. Successful long-term engraftment has been made possible by new drugs that selectively inhibit cytokine production without completely eliminating host selectively inhibit cytokine production without completely eliminating host defense. Conversely, the availability of recombinant cytokines (both natural and modified) has led to great advances in bone marrow transplantation and, more recently, stem cell transplantation.

Cytokine therapy may be used in a variety of ways for bone marrow transplantation. For example, administration of colony-stimulating factors and other hematopoietic cytokines (particularly in combination) can be used to enhance repopulation of bone marrow following chemotherapy or radiation treatment and thus lessening potential morbidity from infections disease or coagulation dysfunction. Alternatively, cytokines may be administered before therapy to enhance the success of autologous transplantation—particularly of stem cells. This expression of stem cells is usually carried out in vivo, although studies are under way to use ex vivo expansion of cell populations for transplant.

### *Infectious diseases*

Resistance to infectious organisms is primary function of the immune system and therefore requires the precise interplay of most of the cytokine catalog. However, the complex and redundant mechanisms of natural and acquired host resistance are probably not amenable to simplistic addition or deletion of individual cytokines. Improvements in the control of infectious disease will more likely involve cytokines as a component of the therapeutic regimes. One promising avenue is the use of cytokines as adjuvants. Although not as powerful as some of the best experimental adjuvants, cytokines have been shown to be at least equal, if not superior, to adjuvants currently approved for human use. Much work remains to be done to optimize this approach; some initial attempts have included using combinations of cytokines, increasing exposure time, and physically associating cytokines with antigens.

Another use of cytokines in treatment and prevention of infectious disease is to correct defects in immune function and thus to enhance host resistance to opportunistic infections common in immunocompromised hosts. This approach usually entails the use of colony-stimulating factors to restore depleted bone marrow, although other cytokines such as IL-1, IL-6, and IFN-$\gamma$ display potential benefits. A more ambitious possibility is the manipulation of cytokine levels (along with other variables) to direct the immune response selectively to either a Type 1 or Type 2 response. At present this approach remains experimental.

## Specific Therapeutic Options

### Exogenous Cytokine Therapy

#### *"Natural" Cytokines*

The discovery of many of the earliest known cytokines took place in human or animal cell cultures, and elucidation of their in vitro an in vivo effects utilized culture supernatants of stimulated primary lymphoid or myeloid cells. These cultures contain a diversity of cytokines that are produced naturally in response to cellular activation. Although these "natural" (contrasted with man-made) cytokines obviously produce desired biological effects, they have not been pursued as intensively as recombinant cytokines for several reasons. First, mammalian cell cultures does not routinely produce cytokines as economically as bacterial fermentations, and mammalian cells are technically more difficult to work with than bacteria or other biotechnology workhorses such as insect cells. However, a more important consideration may be the undefined nature of these cells—particularly the primary cell cultures.

These culture supernatants often contain a multitude of cytokines and other bioactive molecules. This undefined nature makes it difficult to determine the activity of individual factors, which is an undesirable circumstances for conventional drug discovery efforts. On the other hand, a teleological assessment would suggest that just such an assortment of cytokines is perhaps the best application of cytokines and that the cellular source of the cytokine mix represents the ultimate best judge of an appropriate "mix". On the basis of this logic, a mix of natural cytokines may represent an extension of the body's own healing process. One such preparation that has demonstrated clinical efficacy is Leukocyte Interleukin Inj., which has the trade name Multikine.

Multikine is a serum-free lymphokine mixture (containing, among others, IL-2) prepared from human buffy-coat mononuclear cells. Multikine appears to enhance the function of natural killer cells and cytotoxic T lymphocytes. Injection of this product into human with head and neck cancer was found to result in tumor regression with infiltration of the tumors by lymphocytes. Products such as Multikine may hold promise as adjuncts to lymphokine-activated killer cell therapy.

### *Recombinant natural sequence cytokines*

At present, the overwhelming majority of preclinical and clinical studies of exogenous cytokine therapy have utilized natural sequences produced recombinantly. The literature on the results of these studies is therefore extensive, and a recounting of their results is well beyond the scope of this chapter.

### *Mutant or synthetic cytokines*

As described later in this chapter, toxicity is often the limiting factor in protocols involving cytokine administration. Seeking to broaden the therapeutic index of these agents, some investigators have found that rational protein drug design can result in cytokines with enhanced biological activity and reduced "changes in quantitative parameters". This latter phrase was proposed by Brouckaert et al. (1994) as an alternative to "Toxic side effects" in describing the sequelae of exogenous TNF administration and implies a belief that the current state of knowledge does not allow for a precise understanding of whether observed affects are truly toxic or simply adaptive changes. This semantic device is undoubtedly cold comfort to patients enduring the side effects of cytokine therapy (see the section on toxicity associated with cytokine therapy). The majority of published work on cytokine mutant (proteins (muteins) has focused on synthetic versions of TNF.

Tumor necrosis factor is a potent, pleiotropic cytokine; because it is involved in a multitude of both normal an pathological processes, it represents an obvious target for cytokine therapy. However, this very diversity of actions has proven to be problematic when TNF is administered systemically, which often results in profound toxicity. This has led to efforts to design TNF molecules that retain beneficial activity but minimize the toxic side effects. Several investigators have reported success in this effort, although clinical data are not yet available. Another recent example of this approach is the construction of synthetic cytokine (Synthokine) SC-55494, a potent IL-3 receptor antagonist first described by Thomas et al. (1995).

By employing oligonucleotide-directed mutagenesis on a synthetic, natural sequence IL-3 DNA followed by iterative combination, these investigators produced several thousand IL-3 mutants that were subsequently screened for biological activity. The results of this screening identified SC-55494 (Synthokine), a molecule with 48 amino acid sequence changes relative to natural

IL-3. Synthokine was found to produce a ten- to twenty fold increase in hematopoietic activity, although its potential for induction of inflammatory responses was only about twofold greater than natural sequence IL-3.

Subsequent primate studies demonstrated that administration of Synthokine, either alone or in combination with G-CSF, significantly enhance recovery from both neutropenia and thrombocytopenia following radiation-induced bone marrow aplasia. These results suggest that Synthokine and related molecules may hold significant promise in the support of myelosuppressed patients. Several other cytokine muteins have been constructed and are under evaluation for the treatment of various conditions. Examples include IL-4 muteins and IL-5 mutein for the treatment of allergic diseases.

***Fusion proteins***

A different form of cytokine is the fusion protein. The concept behind fusion proteins is to connect the functional portion of cytokine (i.e. the receptor-binding moiety) with another molecule; this second molecule can have biological activity of its own or may be there to assist the function of the cytokine molecule. In the case of fusion proteins consisting of two different cytokines, the resulting molecule is termed a hybrid cytokine or hybrikine. One of the first examples of hybrikine was PIXY321 (Pixykine), which consists of the active domains of IL-3 and GM-CSF coupled by a flexible amino acid linker sequence. By incorporating elements of both IL-3 and GM-CSF (both of which are potent colony-stimulating factors), PIXY321 was found to stimulate a wide range of progenitor cells. Perhaps more significantly, PIXY321 was found to be more active than either of its constituent cytokines, either alone or in combination. Subsequent clinical trials with PIXY321 have shown it to be a useful adjunct in autologous bone marrow transplantation among other indications.

Other cytokine-cytokine fusion proteins are now being described such as CH925, which is a fusion protein comprising rhIL-6 and rhIL-2. This hybrid molecule exhibits erythropoietin-like activity. Cytokines may also be combined with other molecules. For example, construction of cytokine-antibody hybrids produces fusion proteins retaining the properties of both parents (i.e., antigen binding and cellular activation). Such constructs display promise as potent immunostimulants or as treatments for conditions such as septic shock. A different but related immunostimulant property can be obtained by fusing cytokines directly to antigens, as demonstrated by Kim et al. (1997). An interesting example of a novel problem that was overcome by constructing a mutant cytokine was reported by Anderson et al. (1997). These investigators worked with IL-2, a cytokine that has been somewhat problematic to produce recombinantly owing to its heterodimeric structure and because the distinct subunits of the molecule are produced by separate genes on different chromosomes. This problem has been overcome by constructing a single-chain fusion protein (Flexi-12) that retains the biological activity of the natural cytokine molecule. No doubt this approach will find greater utility in the future.

**Methods for Delivering Cytokines**

As previously mentioned, cytokines are proteins or small polypeptide molecules; therefore, oral administration has been impractical because these molecules cannot withstand the digestive environment. Their administration thus for has been primarily parenteral—mainly by intravenous injection. Although certainly the most convenient method clinically, this approach is associated with many serious disadvantages, including the necessity of bolus administration (or at least short-term infusion), which includes toxicities, as well as the need to administer high doses of cytokines to yield sufficient locally effective concentrations (i.e., lack of targeting). A variety of advanced methodologies are under development for a more efficient and clinically manageable delivery system for therapeutic cytokines. Although many of these techniques are in the developmental phase, they show exciting promise as future treatment options.

***Cytokine gene transfer***

One type of gene transfer is designed specificity for treating neoplastic disease. In this approach, cytokine genes are inserted directly into tumor cells, which subsequently produce cytokines. Cytokines exhibit several properties that make them good candidates for treatment of neoplasia, including immunostimulatory activity, the ability to enhance tumor antigen presentation or antigenicity, and direct cytotoxic activity. However, the toxicity associated with systemic cytokine treatment is a major drawback. On the other hand, a more localized administration might lessen this toxicity. In practice, tumor cells from a patient would be removed, specific cytokine genes would be inserted, and these cells would then be reinfused. The theory behind this technique is that the tumor cells will secrete cytokines in vivo; when these cells are encountered by immune effector cells, the secreted cytokines will enhance the antitumor effect, and the tumor cells will be eliminated. Although some success has been demonstrated in experimental animal systems, some cytokines actually appear to enhance tumor growth rather than suppress it. Another potential problem is that gene transfer may induce growth autonomy in the tumor cells and actually exacerbate the problem. Clearly, many issues remain to be addressed before implementing this technique in human therapy.

## Cytokine Inhibition Therapy

### *Chemical (nonbiological) agents inhibiting production and action*

A variety of nonbiological drugs have been found to suppress cytokine production either specifically or nonspecifically. These drugs act via myriad mechanisms, including alteration in cytokine gene transcription or mRNA translation, inhibition of cytokine release, or inhibition of cytokine processing. To date, the primary clinical use of these agents has been as adjuncts to organ and tissue transplantation, in which they have provided great advances in clinical success. Increasingly, these agents are now being evaluated for treatment of other immune-related diseases.

*Glucocorticoids* are well known and powerful immunosuppressive drugs and enjoy wide clinical use for a variety of indications in which inappropriate immune reactions are factors in disease etiology. These drugs act at a variety of molecular targets; germane to the present discussionis their role in modulating cytokine production. Glucocorticoids affect cytokine production via a number of mechanisms including transcriptional repression, posttranscriptional alterations, induction of the suppressive cytokine TGF-β, antagonisms of transcription factors, and cooperation with transcription factors. Glucocorticoids also alter the expression of cytokine receptors. Owing to their highly potent anti-inflammatory and immunosuppressive effects, these drugs are associated with numerous side effects and toxicity.

*Cyclosporin* (*SandImmune*) is probably the most thoroughly studies of the immunosuppressive drugs that specifically target cytokine production. Cyclosporin is a fungal metabolite with several unique features, including a novel amino acid in its composition.

Cyclosporin acts primarily at the level of the T lymphocyte and effectively and reversibly inhibits the action of these cells. The drug exerts this highly specific effect by binding to the cytosolic protein cyclophilin, which is a member of a class of molecules generically known as immunophilins. The cyclosporin-cyclophilin complex targets calcineurin and essentially binds calcineurin to the immunophilin. The calcineurin is unable to interact with downstream transcription factors such as NF-AT, which ultimately prevents the transcription of cytokine genes, including IL-2, IL-3, IL-4, IL-5, and TNF.

*Tacrolimus* (*FK506*) is another immunosuppressive fungal metabolite. Like cyclosporin, FK506 seems to exert its principal effects by altering the expression of cytokine genes—in particular the gene for IL-2. This agent shares little structural similarity with cyclosporin, although it has a similar mechanism of action. Like cyclosporin, FK506 acts by binding to a cytosolic immunophilin; unlike cyclosporin, for FK506 this immunophilin is termed FK506 binding protein (FKBP). Inhibition of IL-2 gene transcription then occurs in a molecular manner similar to that following treatment with cyclosporin.

*Rapamycin* (*Rapamune*, *Sirolimus*) shares structural similarities, including identical binding domains, with FK506; like FK506, it also binds to the cyclophilin FKBP. However, unlike its molecular cousin, rapamycin appears to have only limited (or no) effect on cytokine production but rather appears to affect cell cycle progression in late G1 phase by inhibiting growth factor signal transduction pathways. Although rapamycin does not appear to affect cytokine production, there is some evidence that it may decrease the stability of cytokine RNA within the cell. Work on elucidating the precise mechanism of this drug is under way.

*Leflunomide* (*HWA 486*) is a synthetic immunomodulatory drug (an isoxazole derivative) that has been demonstrated to be effective in animal models of autoimmunity and transplantation rejection. Early reports on its mechanism of action were conflicting, although it was recognized at the outset that the drug operates via different mechanisms than the macrolides described above. In particular, early studies provided conflicting data regarding its ability to suppress cytokine production or the expression of cytokine receptors. More recently, Cao et al. (1996) have demonstrated that leflunomide does in fact work via modulation of cytokine function and thus enhances the production of the immunosuppressive cytokine TGF-β. It is clear that more work is required to identify the drug's exact mechanism of action.

*Thalidomide* has languished for many years owing to its teratogenic effects following its use in pregnant humans. In recent years, however, this drug has demonstrated potential as a useful inhibitor of cytokine production and thus may be enjoying a comeback. For several years thalidomide has been known to be a potent suppressor of TNF production and has recently been demonstrated to suppress production of the immunoregulatory cytokine IL-12. The mechanism of cytokine suppression by thalidomide is currently unknown, although it may act as an immunomodulator (but not necessarily an immunosuppressant) via selective gene regulation. Thalidomide is currently being evaluated for clinical treatment of several conditions such as autoimmune disease and graft-versus-host (GVH) disease.

*Pentoxifylline* is a methylxanthine drug that has reproducibly been shown to inhibit the production of TNF and IL-12 and to affect the production of other cytokines such as IL-1, IL-6, IL-8, and IL-10 variably. Pentoxifylline is showing some promise in maintenance of AIDS patients, although several other clinical conditions may be candidates for treatment with this drug.

*Pentamidine* is an antiprotozoal drug used primarily to treat pneumonia caused by *Pneumocystis carinii* as well as certain inflammatory conditions. In addition to its antiprotozoal activity, pentamidine has been demonstrated to inhibit the production of IL-1 via a posttranslational event—possibly altering the cleavage of the precursor form of the cytokine. More recently, pentamidine has been shown to inhibit the production of inflammatory chemokines.

*Tenidap* is an antirheumatic drug that combines cyclooxygenase inhibition with suppression of the acute phase response. Tenidap has been demonstrated to suppress the production of IL-1, IL-6, and IFN-γ. It appears specifically to suppress cytokine production by altering ionic homeostasis, although the exact mechanism of action remains unknown.

Other drugs have also been shown to inhibit cytokine production. The precise mechanism of action of many of these drugs is currently unknown, and they have not been developed extensively for clinical use as specific cytokine inhibitors. As knowledge of the molecular biology of cellular activation grows, drugs will undoubtedly be formulated to modulate specific mechanisms.

### ***Biological agents that inhibit cytokine production or action***

#### *Specific inhibitory cytokines*

To date, only one naturally occurring cytokine antagonist has been identified, namely interleukin-1 receptor antagonist (IL-1RA), which exists in two forms: secreted and intracellular. This antagonist is a member of the IL-1 cytokine family (which includes IL1α and IL-β) according to amino acid sequence, receptor binding avidity, and gene structure and location. This agent seems to function as a pure receptor antagonist and does not exhibit any discernible agonist activity (e.g., internalization of receptor complexes, etc).

Experimental evidence suggests that IL-1RA plays a role in various disease states such as rheumatoid arthritis (RA), sepsis, diabetes, and other diseases. Understandably, the clinical application of a natural receptor antagonist represents an ideal clinical tool; clinical trials are under way for treatment of RA. Mutant proteins have also been constructed that act as a specific receptor antagonist for both IL-4 and IL-13 activity. This early success should lead to the creation of additional specific cytokine antagonists that are therapeutic.

#### *Cytokine receptors*

Although most cytokines appear to exert their biological activity via interaction with specific cell-surface-bound receptors, many cytokine receptors are known to be present in a soluble form in the circulation of normal, healthy individuals. These soluble receptors may bind their cognate antigen with the same affinity as the surface receptors and have been postulated to have various physiological functions. One possible function is to serve as a chaperone molecule protecting the cytokines as they are ferried though the circulation. Another possible function, and one that serves as the basis for their possible use as therapeutics, is a binding molecule for free cytokines in the circulation, which serves to control their bioavailability.

Most of the work in this area has involved the IL-2 receptor, although several other soluble receptors have been investigated as well. A potential problem (or benefit, depending on the desired outcome) inherent in the use of soluble cytokine receptors is their propensity for actually enhancing the activity of certain cytokines—possibly by extending their presence in the circulation. Pharmacokinetic studies with recombinant cytokine receptors have demonstrated variability in the clearance between the different receptors as well as with different forms of the same receptor. Given the structural and functional differences between the different cytokine receptor subfamilies, this is perhaps not very surprising. However, this finding does point to the need to evaluate these molecules on a case-by-case basis in their development as human therapeutics.

*Anticytokine antibodies*

Like soluble cytokine receptors, naturally occurring anticytokine antibodies have been demonstrated in the circulation of normal healthy individuals. Also, like the soluble receptors, these antibodies are thought to represent a mechanism for controlling the bioactivity of cytokines and to prevent them from inducing inappropriate responses. These molecules have been invaluable research reagents in understanding the role of cytokines in myriad as therapeutics. Several anticytokine antibodies (particularly anti-IL-4, and anti-IL-5 antibodies) have been evaluated with some success in the treatment of allergic disease. A variety of other conditions have been treated with anti-cytokine antibodies with varying degrees of success. As with the addition of exogenous cytokines, treatment of disease with single anticytokines antibodies may be confounded by the highly redundant nature of cytokines.

*Fusion toxins, immunotoxins and chimera toxins*

In a slightly different use of anticytokine biotherapeutics, the cytokine-producing cell rather than the cytokine itself is targeted. As discussed elsewhere, the overproduction of cytokines, or at least an overly vigorous response by cytokine receptor-bearing cells (e.g. lymphocytes), can lead to a number of pathological conditions such as autoimmunity or neoplasia. The greatest degree of specificity could be achieved by developing a so-called "magic bullet," a molecule that is both highly specific and toxic. Potential candidates for this role are fusion toxins, which are also termed immunotoxins or chimera toxins. Numerous highly potent toxins are found in nature, and three of these in particular have been used in the construction of immunotoxins: diphtheria toxin and *Pseudomonas* exotoxin (both bacterial products) and ricin (a plant product). These molecules share several structural features, including a binding domain allowing them to attach to cells, a component that allows the molecule to cross the target cell membrane, and an enzymatically active domain (the toxophore) that poisons the cell. Unfortunately, the binding domain allows the toxins to attach to many different cell types. However, by removing this binding domain and replacing it with a portion of the cytokine or growth factor of interest the toxicity of the resulting molecule may be tailored to specific targets.

The resulting hybrid molecule is now specific only for cells bearing a particular receptor; the molecular components mediating translocation and intoxication remain functional. Two of the earliest successes with this approach were the toxins $DAB_{486}$-IL-2 and $DAB_{389}$-IL-2, which contain fragmentary portions of the IL-2 molecule. Clinical trials and experimental studies with these immunotoxins have found them to be relatively safe and effective for a wide variety of conditions, including inhibition of HIV-1 RNA, mycosis fungoides, rheumatoid arthritis, and IL-2 receptor-expressing malignancies.

**Table 4.2. Various cytokine fusion toxins**

| *Toxin* | *Cytokine* |
|---|---|
| Diphtheria toxin | G-CSF |
| | IL-6 |
| | IL-15 |
| | GM-CSF |
| | IL-3 |
| Ricin | IL-1 |
| | GM-CSF |
| Pseudomonas exotoxin | IL-4 |
| | IL-6 |

Several other fusion toxins have been constructed. Most of these novel toxins are still experimental, and thus their actual utility as therapeutics remains to be established.

*Antisense technology*

With a more complete understanding of molecular genetics has come the ability to alter fundamental biological processes experimentally and therapeutically. Nowhere is this manifested more than in the emerging field of antisense technology. Antisense technology is based on the construction of synthetic oligonucleotide sequences that will bind specifically with "sense" (i.e., message sequence) nucleic acids and in effect mask them and prevent sequence-specific actions such as translation of mRNA into protein.

In theory, antisense therapy could represent a highly specific tool able to interrupt undesired biological processes at their most fundamental level. The advantages relative to cytokine therapy are obvious, for antisense technology allows a highly selective shutdown of specific cytokines or related growth factors to occur while theoretically leaving other factors unaffected. This would make antisense more desirable than even fusion because it involve deactivation—rather than destruction—of a cell. Although concept of antisense technology has been demonstrated repeatedly in the laboratory, many important problems remain to be solved before this technique becomes a routine therapeutic option. These include knowledge of the proper sequence, target accessibility, specificity, and drug stability and delivery.

One potential anticytokine use that appears to have bypassed this last problem is the administration of antisense oligonucleotides topically as an adjunct to wound healing. Another and more serious problem with antisense therapy has been unanticipated toxicity—particularly cardiovascular changes. At present, antisense technology appears to be more valuable as a research tool for understanding cytokine biology than as a component of cytokine therapy.

*Extracorporeal removal of cytokines*

In situations where in situ inactivation of cytokines may be impractical or impossible, a more brute force option may be the physical removal of cytokines. One approach that has been investigated is removal of cytokines during continuous hemofiltration. To date, the result of this approach have been equivocal; although removal of certain proinflammatory cytokines from the filtrate has been demonstrated, there has been little evidence of a reduction in systemic cytokine levels.

Furthermore, improvements in patient outcome do not appear to be clinically relevant. Further clinical trials of this approach are under way. A more selective approach for physical removal of cytokines from the circulation is by extracorporeal binding to antibodies. This technique was reported by Weber and Falkenhagen (1996) and was termed the Microspheres Based Detoxification System. In essence, polyclonal antibodies specific for IL-1, IL-6, and TNF were covalently linked to microspheres, and human plasma spiked with these cytokines was processed through the system.

The authors reported efficient removal of these cytokines using this technique, for the rates of removal were greater than those reported following ultrafiltration. This technique has at least two advantages over removal by hemofiltration. First, the process allows for a highly selective removal of targeted cytokines rather than bulk removal of molecules. Second, antibodies to other molecules can also be attached concomitant with the anticytokine antibodies; this might facilitate the removal of other molecules. Second, antibodies to other molecules can also be attached concomitant with the anticytokine antibodies; this might facilitate the removal of other molecules associated with a particular disease state such as endotoxin during sepsis. The clinical effectiveness of this strategy has yet to be demonstrated.

## Monitoring the Results of Cytokine Therapy: Some Considerations

With the capability of altering the balance of cytokines in vivo the accurate determination of the level of these molecules naturally becomes of great importance. Cytokines are routinely measured by one method or a combination of three methods: bioassay, immunoassay, and molecular assay. Although each of these assay types has its own spectrum of advantages and disadvantages, an inclusive comparison is beyond the scope of this review. Nonetheless, by far the most commonly employed type is the enzyme-linked immunosorbent assay (ELISA), which is fairly sensitive (and with certain modifications, highly sensitive), highly specific, relatively inexpensive to perform, an technically straightforward. Also, ELISAs are easily adapted to automation, which is an obvious advantage in the clinical laboratory. Given their current and almost certain future acceptance, it is prudent to take a closer look at the potential problems and technical issues associated with the use of ELISA for quantitating cytokine levels in clinical samples. Examples of some of the more critical considerations are as follows:

1. Effect of sample processing: Several studies have demonstrated that variables in sample collection, processing, and storage can affect the result of cytokine assays.
2. Existence of alternative molecular forms of the antigen: Certain cytokines can exist in variant forms (e.g. IL-6) or as precursor

molecules (e.g., IL-1β). The type of antibody used in the ELISA becomes an issue in these circumstances in that monoclonal antibodies (often used in commercial ELISAs) may not necessarily detect the presence of such alternative forms.

3. Presence of interfering substances: A wide variety of substances may be present in clinical samples and can interfere with immunoassays. Some of these include soluble receptors, natural antagonists such as IL-1RA, chaperone molecule such as macroglobulin, and cytokine-specific autoantibodies.
4. Assay precision: Although excellent reagents and kits are increasingly available, assay precision is still an important consideration in cytokine measurement. At present the most reliable method for increasing precision is strict attention to methodological details.
5. Reference standards: A variable that is receiving increased attention is the source of the reference material used in the assay. Because experiments are ultimately extrapolated back to the reference preparation, this represents a weak link in the assay. However, until recently there has been limited effort to standardize the reference among commercial suppliers of cytokine ELISAs; thus, results obtained with any given kit may not match those obtained with other kits. It is to be hoped that this essential item of quality control will soon be addressed.

Unlike many other biological entities (e.g., hormones. enzymes, etc.) in which normal human clinical ranges have been established, the expected or "normal" values for cytokines in various biological samples have not been published. A comprehensive review of the literature suggests that the vast majority of such evaluations have been made to compare cytokine levels present in humans during particular disease states with normal (i.e., appropriately matched control) individuals.

A secondary source of such information is the product inserts for various commercial immunoassays kits that quote the results of internal studies performed by the supplying company. Unfortunately, an exhaustive compilation of all available data is still lacking, although such a future compilation will be of great value. Even if some day technical assays are perfected and comprehensive baseline values are compiled, a final consideration must be made; namely, which cytokines should be evaluated and under what circumstances? For example, monitoring the levels of an exogenously administered cytokine for pharmacokinetics would be relatively straightforward—particularly if a novel cytokine were administered (e.g., mutant cytokines with a unique protein structure), and might be useful for maintaining "therapeutic levels" (if in fact such levels have been established). On the hand, because baseline cytokine levels are low or undetectable, measuring these factors in the circulation would be essentially useless if one were evaluating potential immunosuppression (either as a disease sequela or clinically induced). In such cases, measurement of ex vivo cytokine production by stimulated cells would be required.

A final consideration concerns the evaluation of cytokine-related biomarkers. For example, neopterin, a biologically stable molecule induced by interferon-gamma, might be used to monitor therapy with this molecule. Conversely, certain cytokines such as $\alpha_2$-macroglobulin are known to interact with cytokines. Although these molecules do not appear to affect the function or bioavailability of cytokines significantly, more information is needed to assess their contribution fully.

## Toxicity Associated with Cytokine Therapy

Preclinical toxicology is an integral and vital component of all pharmaceutical and biotechnological drug development. In toxicological evaluation, the potential of a drug candidate to affect various organ systems adversely at up to super physiological doses, is estimated to establish a margin of safety before the inhibition of clinical trials in humans. With the advent of genetically engineered proteins in recent decades, an era of more "natural" forms of therapy was anticipated.

From the early (and, in retrospect,, perhaps a bit naive) assumption that "if a little is good, then more is better" view of cytokines and other biological therapeutics, a realization has emerged that the road to rational biotherapy will be somewhat more difficult. At the risk of fatal oversimplification, the preclinical and clinical evaluation of cytokines comes down to two points: (1) traditional animal toxicology models may not reproducibly predict the ultimate human toxicity or recombinant molecules, particularly cytokines, and (2) the sequelae of in vivo cytokine administration (and perhaps cytokine suppression as well) can not necessarily be predicted on the basis of their in vitro actions. Let us examine these points in greater detail. The first problem, namely the inability of routine animal models to predict the toxicity of cytokines, can best be understood by considered the basic biology of the cytokines themselves. Three essential features are of particular importance:

1. Certain cytokines (e.g., IL-3, IL-4, IL-12, GM-CSF, and IFN-γ) exhibit a relatively high degree of species specificity. Unlike most nonbiological therapeutics, cytokines mediate their normal physiological (and apparently at least some of their deleterious) effects via their interaction with highly specific receptor molecules. This receptors interaction is also the basis for the strict species specificity of certain cytokines. Thus, administration of such molecules to an inappropriate species may be expected to result in difficulties in interpretation.
2. Although many recombinant proteins are highly homologous to their endogenous counterparts, minor biochemical discrepancies (e.g., base pair substitution, alternative forms of glycosylation, etc.) may result in altered biological function or, more likely, antigenic stimulation and lead to the induction of neutralizing antibodies. Although not as common when cytokines are evaluated in a homologous species (e.g. mouse-mouse), when human sequence cytokines are injected into laboratory animals, neutralization of the protein's bioactivity will result within a few weeks. Consequently, effects that may not be apparent in humans except after chronic administration will effectively be masked in animal models.
3. Whereas structure and function have been relatively well conserved for most mammalian cytokines, certain to these molecules display divergent function between species. In such cases, the clinical sequelae of human cytokine therapy may not be predictable even from appropriate designed animal studies.

In spite of these potential difficulties, progress is being made in preclinical assessment of cytokine toxicity. For example, transgenic animals have been created that have a high constitutive level of cytokine production and thus mimic the effect of super-physiological doses of cytokines in humans. Another approach is the construction of transgenic animals expressing human cytokine receptors. The second problem, namely that the clinical response to in vivo cytokine administration is not necessarily predictable by in vitro data, is likewise a complex issue. Let us once again examine his in the light of basic cytokine biology:

1. As stated earlier, cytokines are pleiotropic and redundant in function. An example of this is provided by tumor necrosis factor (TNF), which exhibits potent cytotoxicity for isolated (i.e., in vitro) tumors cells and induces the regression of solid tumors following in situ injection. However, TNF's broad diversity of physiological actions (not the least of which is the mediation of systemic shock) precludes its use as a systemic therapeutic. Cytokine redundancy may account for certain toxicities by inappropriately enhancing (or suppressing) the effects of other endogenous cytokines.
2. Cytokines appear to function principally as carriers of bioinformation at a local level. Although cytokines are now known to mediate the functions of the immune, nervous, and endocrine systems (the neuroimmunoendocrine axis), most of these interactions probably take place at discrete sites. This impression is supported by the demonstration that the systemic (circulating) levels of almost all cytokines are extremely low and that exogenously administrated cytokines are rapidly cleared from the circulation.

A common unintended consequence of cytokine treatment—although not a toxicity per se—is the development of anticytokine antibodies. This may seem surprising intuitively, for cytokines are endogenous molecules. However, anticytokine antibodies have been demonstrated in the sera of normal healthy individuals and are now thought to constitute part a complex regulatory system for controlling the activities of endogenous cytokines. Thus, it should not be surprising that administration of superphysiological concentrations of these proteins induces the production of specific antibodies. Moreover, as mentioned elsewhere regarding animal studies, even recombinant proteins may not duplicate the natural product, which results in the creation of novel antigens and the subsequent development of antibodies. Anticytokine antibodies may be either binding or neutralizing.

Naturally, of the two, neutralizing antibodies represent the greatest concern because their presence could effectively negate the therapeutic benefit of treatment. Owing to the number of studies evaluating these molecules, most of the neutralizing antibody information has been gained from clinical trials of interferons and interleukin-2. However, these are not isolated incidents, and minimizing the development of anticytokine antibodies will be an ongoing challenge to clinicians. Aside from the issue of neutralizing antibody formation, administration of almost all exogenous cytokines tested to data has been associated with a range of toxicities of various severity. One toxic side effect that appears to be common to a number of different cytokines (particularly IL-2 and GM-CSF) is the so-called vascular leak or capillary leak syndrome. This syndrome is characterized by such symptoms as peripheral and pulmonary edema, hypoxia, occasional ascites, perhaps hypotension, and sometimes respiratory failure. Although the syndrome is suspected to be of immunological origin, the exact mechanism of toxicity remains to be elucidated.

## Conclusions

The concept of harnessing the body's own natural healing powers has long been considered the ideal form of medicine. The traditional avenues available for accomplishing this have taken the form of herbs (whether crude concoctions or semisynthetic formulations) and spiritualism (ranging from magical incantations to prayer). However, in recent years, the discipline of molecular biology has provided researchers and clinicians with the highly precise tools to make this promise a reality at last.

Genetic engineering allows us to produce products is essentially unlimited amounts, thus making it possible to replace missing biochemical where a deficiency exist. Likewise, technologies such as antisense and hybridoma-produced monoclonal antibodies allows us to deplete "unwanted" biomolecules selectively. Nevertheless, as pointed out in this chapter, a comprehensive understanding of the design of the vertebrate organism still eludes man. The myriad interconnected and interdependent biological processes work together in health; it may be the height of naivete to assume that the simple addition or deletion of a single molecule will always effect a cure, disease such as diabetes notwithstanding.

As is being reaffirmed in clinical studies, such is the case with cytokines. Not that success is completely elusive. For selected applications such as bone marrow reconstruction, organ and tissue transplant, and several immunological disorders, cytokines are providing to be an exciting addition to the medical armamentarium. Moreover, certain innovations have resulted in "better than nature" molecules that can be tailored to specific pathologies, and the future promises even more improvements. It is only reasonable to temper our enthusiasm with the knowledge that much remains to be learned.

# Chapter 5

# Food Allergens

As discussed already, new plant varieties that are potential food sources are being, developed with recombinant DNA technologies. On the basis of these techniques, specific genetic modifications that could not have been accomplished through traditional breeding techniques have been achieved, including the expression of proteins into plants from other species. Several potential foods that have been derived from these new plant varieties by recombinant DNA technologies and are either commercially available or approaching commercial introduction. Because genes governing these new traits code for proteins that are ordinarily not present in that plant, there is concern about the effects of the transferred proteins on individuals ingesting them. A major concern is the potential of these transferred proteins as food allergens.

## Food Allergy

An adverse reaction to food is any clinically abnormal response attributed to exposure to food or food additive; these reactions include those of those of both immunologic and nonimmunologic pathogenesis. Food allergy (food hypersensitivity) is a specific type of immunologically mediated adverse reaction resulting from the ingestion of a food or food additive. This reaction occurs only in some of the exposed subjects, requires prior exposure (sensitization), may occur only after a small amount of substance is ingested, and is unrelated to any physiological effect of the food or food additive.

Most food allergies are mediated by IgE antibodies and have a rapid onset—typically within minutes of exposure. Other adverse reactions to foods are food intolerance, which is an abnormal physiologic response to an ingested food or food additive; food poisoning, which is basically a toxic reaction; and pharmacologic reaction to foods, which are due to chemicals present in food that produce a drug-like effect. These last three reactions are nonimmunologic. These reactions may be confused with true food allergy because they have similar symptoms. In the induction of an allergic response, proteins that are typically innocuous enter the body via mucosal surfaces and are processed by local antigen-presenting cells that present peptide fragments to the allergen to specific cells called T lymphocytes or T cells. Recognition between the TH2 lymphocyte (a subset of T lymphocytes) and antigen-specific B lymphocytes (the cell that eventually produces antibody) results in the release of soluble factors, including interleukins 4, 5, and 13; these, in turn, stimulate B-cell proliferation and differentiation and the production of allergen-specific IgE.

There is some evidence for genetic control of IgE antibody production to specific allergens. The portion of the protein molecule that interact with T cells are called T-cell epitopes and those that interact with B cells or antibodies are called B-cell epitopes. The IgE antibodies have the ability to bind to specific receptors on mast cells or basophils, thus sensitizing these cells. Subsequently, when an allergen reaches the sensitized mast cell, it cross-links surface-bound IgE, triggering the release of preformed and newly synthesized mediators. These mediators, in turn, elicit the clinical

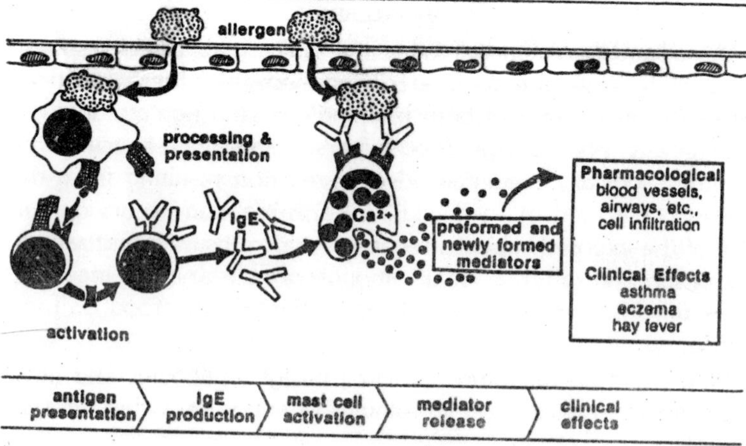


*Fig. 5.1. Sensitization to food allergens.*

signs and symptoms of allergic diseases, including asthma, eczema, urticaria, hay, fever and anaphylaxis. Precise figures concerning the prevalence of food allergies are difficult to obtain.

In clinical surveys it has been estimated that about 500,000 to almost 1 million children younger than 6 years old have reproducible allergic reactions to foods. Studies in adults suggest that almost 2 to 4 million adult Americans are sensitive to foods or food additives. These figures contrast dramatically with the public's perception of the importance of allergic reactions to foods. It has been reported that at least one in four atopic adults believe that they have experienced adverse reactions following the ingestion or handling of foods. Similarly, parents believe that one of four of their children has experienced at least one adverse reaction to a food.

### Factors that Affect the Development of Food Allergy

The quantity of a particular food ingested, which is influenced by the diet and culture of a region, can have a significant effect on the prevalence of a specific food allergy. The high prevalence of codfish allergy in Norway, rice and soy allergy in Japan, and peanut allergy in the United States are good examples. Route of exposure is another important factor in the development of food allergy. Not only are food allergens ingested, but exposure can also occur through skin contact and inhalation. Gut permeability has long been thought to be a major factor in food allergy sensitization, but there has been very little investigation of this subject. Processing of food is another factor that can substantially alter food allergen content. Finally, intrinsic factors are clearly present; some foods apparently are very allergenic, whereas other foods are rarely if ever allergenic. This is discussed more fully in the next section on common food allergens that follows.

## Common Food Allergens

### Food Allergens

Most foods contain literally thousands of different proteins to which and individual could potentially react. What makes some proteins within a particular food allergenic is not understood. Generally, allergens, including food allergens, can be classified as either major or minor. Major allergenes are those to which 50% or more of a population sensitized to that particular food react. Reactions can be defined by skin test for IgE antibody response;

however, more recently major or minor allergens are defined by reactivity to the individual allergen following testing of the sensitized subjects' sera by immunoblot reaction. In addition to allergens being major or minor, they can also be isoallergens.

Isoallergens are homologous proteins of similar molecular size, identical biological function, and at least 67% amino acid sequence identity. Isoallergens are homologous proteins that are identical immunologically which results in slightly different isoelectric points. Thus, several immunologically identical spots that are of similar yet distinct isoelectric points or molecular weights may occur during two-dimensional electrophoresis and appear to be distinct proteins. Specific designations may be assigned to thoroughly characterized allergens according to the accepted taxonomic name of their source. The first three letters of the genus plus the first letter of the species name are used to indicate the source.

An arabic number is assigned according to the temporal order of allergen identification; however, generally the same number should be used to designate homologous allergens. For example, the first allergen described in brown shrimp, *Penaeus aztecus*, is designated Pen a 1, and the homologous molecules from Indian shrimp, *P. indicus*, is named Pen i 1. The nomenclature system also provides rules for describing allergen genes, mRNAs. cDNAs, and, most importantly, recombinant and synthetic peptides of allergenic interest. Foods that most commonly elicit allergic reactions are cereal grains, cow's milk, crustacea, fish, hen's eggs, legumes, mollusks, and tree nuts. The following section will briefly review some of the commonly allergenic foods and the allergenic proteins that have been identified from these foods. Generally, these foods have been identified as allergens based on the demonstration of positive skin test reaction or in vitro assay for specific IgE antibodies following a positive history; however, the preferred method of diagnosis is though double-blind, placebo-controlled food challenges (DBPCFC).

## Animal-derived Food Allergens

Although several foods from animal sources have been implicated in allergic reactions, there are only a few major food allergens that have been well studied. Many of these food (milk, eggs, fish crustacea) and their allergens are described in more detail in the following paragraphs. Several milk proteins have been identified as allergenic or immunogenic in humans. Many patients are allergic to more than one milk protein. Individual allergic to cow's milk often have IgE antibodies directed against goat's or sheep's milk. Caseins and β-lactoglobulin appear to be the most frequently involved allergens in cow's milk sensitivity.

Egg allergy is one of the most frequently implicated causes of immediate food-allergic reactions in children in the United States and Europe. Although there is extensive cross-reactivity among proteins from various birds, hen's egg proteins seem to be slightly more allergenic than those of duck eggs. Egg white (albumin) appears to be more allergenic than yolk. Egg-white proteins have been extensively studied; most have been purified and their amino acid sequences determined. Several major egg-white proteins have been identified as major allergens. Ovalbumin (Gal d 2) comprises more than 50% of egg-white protein. Ovotransferrin or conalbumin (Gal d 3) contributes 12% of the total protein of egg white. Ovomucoid (Gal d 1) comprises 10% of egg-white protein. Lysozyme (Gal d 4) is a small 14.3-kDa protein with isoelectric point (pI) of 10.7. It is a single polypeptide chain (with 129 amino acids cross-linked by four disulfide bonds.

In addition to egg-white proteins, yolk proteins may also be allergenic. Apovitellenin I is a 9-kDa protein, and apovitellenin VI is a 170-kDa protein, the two of which comprise 2% of egg-yolk protein. Phosvitin (an iron-carrying molecule comprises 10% of total egg-yolk protein. Livetins are derived from the blood of the hen and can also be found in the yolk. Both phosvitin and livetins are potent allergens in some egg-sensitive individuals. The consumption of fish or inhalation of cooking vapors from fish are frequent causes of IgE-mediated reactions. There have been no published reports on the prevalence of IgE-mediated reactions to a particular species of fish, for most studies refer only to cod or to "fish" in general. However, fish is one of the most commonly implicated allergenic foods and has been incriminated in fatal anaphylactic reactions.

Although the overall prevalence of fish allergy is unknown, the incidence of fish hypersensitivity is observed to be higher in countries where fish consumption is above average. One of the most comprehensive analyses of a food allergen has been the elegant work by Aas, Elsayed, and colleagues that resulted in purification and characterization of the codfish allergen, Gad c1 (originally designated Allergen M) Gad c 1 belongs to a group of muscle tissue proteins known as parvalbumins that control of the flow of $Ca^{2+}$ in and out of cells; parvalbumins are only found in the muscles of amphibians and fish.

The existence of structurally related parvalbumins in different species of fish may explain cross-reactivity in fish-allergic individuals because Gad c 1 shares approximately 34% homology with similar proteins from hake, carp, pike, and whiting. At least 30 edible species of crustacea are commonly consumed in the United States. Crustaceans (phylum Arthropoda, class Crustacea) include shrimp, prawns, crabs, lobsters, and crawfish, which are frequent causes of food hypersensitivity. As is true for fish, a higher incidence of allergy to shellfish is seen in geographical areas where more is consumed on a regular basis. Shrimp is the most studied of the crustacean allergens. Daul et al. (1991, 1992, 1994) first isolated the major shrimp allergen (Pen a 1) from boiled brown shrimp (*P. aztecus*) and identified it as the shrimp muscle protein tropomyosin. Pen a 1 has a molecular weight of 36 kDa and is readily isolated from the boiling water and meat of cooked shrimp. This allergen constitutes about 20% of water-soluble protein in crude cooked shrimp extract and inhibits up to 75% of IgE binding of pooled shrimp-sensitive subjects' serum to whole-body shrimp meat extract.

The allergen-bound IgE in 28/34 (82%) sera from shrimp-sensitive individuals. Pen a 1 is composed of 284 amino acid residues and 2.9% carbohydrate: it has a pI of 5.2 Pen a 1 is the same as Pen i 1, the major allergen isolated from a different species of shrimp—*P. indicus*. Leung et al. (1994) produced a recombinant shrimp allergen from a cDNA library of the greasyback shrimp, *M. enis*. That allergen has 284 amino-acid residues, is similar in amino acid composition to Pen a 1 and Pen i 1, and has a molecular weight of 34 kDa. In immunoblotting studies, recombinant Met e 1 allergen bound IgE in serum samples from all (8/8) individuals in the study with histories of anaphylactic reactions to shrimp. The authors confirmed the observations of other groups identifying the 34-kDa allergen as shrimp tropomyosin.

**Food Allergens of Plant Origin**

Several plant foods are major food allergens. These include legumes, particularly peanuts and soybeans; seeds and nuts; fruits; vegetables; grains; and species. Several important food allergens of plant origin are discussed below.

Peanuts, although very popular, are the most allergenic food known. Peanut allergy is seldom outgrown, and allergic reactions to peanuts are often acute and severe. Peanut proteins have customarily been classified as albumins (water soluble) or globulins (saline soluble). Most peanut storage proteins are globulins; they make up 87% of the total protein. Over the years, peanut proteins have been further fractionated and include albumins, arachin, and conarachin or nonarachin. Globulins include two major proteins, a-arachin and a-conarachin. A multiplicity of peanut allergens have been demonstrated, yet, with the exception of Ara h 1 and Ara h 2, have yet to be fully characterized and identified. Part of the problem in identification of peanut allergens lies in the large number of allergenic peanut proteins.

Burks et al. (1991) identified a 63.5-kDa molecular weight glycoprotein peanut allergen using immunoblotting and enzyme-linked immunosorbent assay (ELISA) method and sera from peanut-sensitive atopic dermatitis patients. This allergen (Ara h 1) has a pI of 4.55. In a later report, a second peanut allergen (Ara h 2) was identified and purified with a molecular weight of 17 kDa and an isoelectric point of 5.2. These studies of Burks and coworkers are the most advanced to date in the characterization of major plant-derived food allergens:

Globulins are also major proteins of the soybean. By adjusting the pH of the saline-soluble soybean protein fraction to 4.5, the globulins precipitate and leave a whey fraction that constitutes 6-8% of the protein. The whey fraction contains haemagglutinin, trypsin inhibitors, and a urease. Several soybean proteins have been studies as allergens. A study of Kunitz soybean trypsin inhibitor (KSTI) as an allergenic protein was prompted by a soy-allergic woman working with KSTI in an occupational setting. However, KSTI appears to be a relatively minor allergen. Herian et al. (1990) described a 20-kDa IgE binding protein from soybean and designated it S-II. Two serum samples from subjects allergic to soy showed IgE binding to a 20-kDa band, whereas no IgE binding was observed to pure KSTI. Roasting of the soybeans appeared to enhance IgE binding to a 20-kDa allergen.

The allergen Gly m 30 K is described by Ogawa et al. (1991) as a 30-kDa-molecular-weight protein that is neither a soybean lectin nor the basic subunit of the 7S globulin but rather is a minor protein of the 7S globulin fraction. Sixty-five percent of the subjects in this study had specific IgE for Gly m Bd 30K; however, these individuals did not experience severe or anaphylactic reactions to soy.

Rice (*Oryza sativa*) is a dietary staple for approximately one-half of the world's population. In japan rice has been found to aggravate atopic dermatitis frequently through IgE-dependent mechanisms. The major rice allergens consist of micro-heterogeneous albumin proteins with molecular weights ranging from 14 to 60 kDa. The allergens are encoded in a multigene family. The nucleotide sequence of a cDNA coding for the major rice allergen has been identified. The mature protein has a molecular weight of approximately 14 kDa.

Tree nuts are major plant food allergens. Brazil nuts are a cause of serious systemic anaphylaxis in some individuals. Arshad et al. (1991) found several allergenic fractions by immunoblotting using serum from allergic individuals to detect Brazil nut protein allergens. Another cause of significant nut-allergic reactions is almond. Bargman et al. (1992) used immunoblotting techniques to detect IgE binding proteins in almond extracts. An extensive number of proteins with molecular weights ranging from 38 to 70 kDa bound IgE. Two major allergens were identified. One is a 70-kDa heat-labile protein, whereas the other is a 45-50 heat-stable protein.

**Cross-Reactivity Among Food Allergens**

Cross-reactivity can occur among food allergens. For example, within a given food group, clinically shrimp-allergic patients report reactivity to other crustacea such as crab, lobster, and crawfish in the absence of prior exposure. Cross-reacting allergens among crustacea have been confirmed by in vitro immunochemical studies. Fewer reports exist of cross-reactivity within vegetable food groups; however, one well-studied plant group is the legumes. Significant cross-reactivity, based on in vitro assays, is observed among peanuts, garden pea, soybeans, and chickpeas.

In contrast to crustacea sensitivity, the clinical hypersensitivity to one legume does not necessarily mean hypersensitivity to all legumes. In addition to cross-reactivity of foods within the same group, cross-reactivity has been descried between foods and substances that are botanically unrelated or only distantly related. For example, it has been reported that ragweed pollen cross-reacts with melons and bananas, grass pollen cross-reacts with celery and a variety of other vegetables and birch pollen cross-reacts with several fruits. Recently it has been shown that latex allergens (newly identified proteins in natural rubber products) cross-react with fruits, including avocados, chestnuts, and bananas. Studies from our laboratory have shown that marine animals belonging to different phyla such as oysters and crustacea or clams and shrimp cross-react. How can proteins form such distantly related substances cross-react?

One explanation may be the presence of common structural or functional proteins. Profilins are ubiquitous cytoskeleton proteins that have a high degree of amino acid sequence homology and similar allergenicity. Kraft and Valenta proposed profilins as a pan allergen. This suggests that molecules of similar structure may have the same allergenicity. Our studies and those of others demonstrate that the principal shrimp allergen is tropomyosin, a major muscle protein. However, in contrast to profilins, tropomyosins in beef, pork, and chicken are not allergenic, although they have at least 60% amino acid sequence homology with shrimp tropomyosin. Thus, one must be careful when inferring cross-reactivity for all common structural proteins or proteins with similar amino acid sequences. The clinical significance of cross-reacting allergens is not entirely clear. However, cross-reacting allergens in different plants and animals poses serious concerns about the cross-reactivity of known allergens with transgenic products.

## General Characteristics of Foods as Allergens

### Physical Properties

Generally, most food allergens are glycoproteins with acidic isoelectric points. This property is true for many antigens; thus it is not a unique property for food allergens. Most known food allergens have molecular weights between 10 and 70 kDa. Although smaller molecules could act as haptens, the molecular weight of 10 kDa probably represents the lower limit for the immunogenic response. The upper limit is probably a result of restricted mucosal absorption of larger molecules. However, some allergens such as Ara h 1 (63.5 kDa) and Ara h 2 (17 kDa) exist in their native forms as parts of large protein polymers that are 200-300 kDa in size. It is not known whether these large molecules act as allergens or must be dissociated during the digestive process.

### *Valancy*

Because all allergens must be able to bridge IgE molecules on the surface of mast cells to cause degranulation, they are constrained in their molecular dimensions. Allergens must contain at least two IgE antibody-reactive sites to trigger mediator release. However, because some monovalent allergens, such as the 21-residue venom peptide mellitin or the ovalbumin-IgE monoclonal antiovalbumin experimental model, can still elicit histamine release from basophils or mast cells or generate anaphylactic reactions in mice, allergens may have fhe ability to bind to surface IgE antibodies on basophils and mast cells and aggregate or aggregate and then bind, thus converting the monovalent allergens to polyligands that can trigger allergic reactions. It is not known whether this occurs in vivo in patients with allergic disease but it is a substantial factor in allergic reactions.

### *Stability and resistance*

Allergens must reach the intestinal tract in an immunologically active form to exert their effects. Thus, conventional wisdom is that these proteins are comparatively resistant to heat or acid treatment as well as proteolysis and digestion; however, important exceptions to exist. Many allergenic food proteins are heat-resistant (i.e. peanuts, shrimp, cow's milk, fish). The application of heat promotes protein denaturation and the loss of conformational epitopes, but the resistance of many food allergens to heat denaturation suggest that conformational epitopes are not always critical for IgE binding. Although heat treatment leads to protein denaturation and loss of conformational epitopes, enzymatic or acidic cleavage of the polypeptide chains may destroy both conformational and linear epitopes. Resistance to digestion and other enzymatic processes are dependent upon the nature of the enzyme or hydrolytic system used in the experiments, the choice of systems for assessing the immunogenicity of the hydrolysis products and, to a lesser extent, the specific food allergen being evaluated.

### *Epitopes*

Our knowledge of T-cell epitopes of food allergens is largely extrapolated from non-food allergens; however, food allergens may have unusual properties. Food allergens are recognized by gut-associated lymphoid tissue (GALT), and the immunoregulatory mechanisms of T-cell responses of this system have not been well defined; thus, there may be unique properties to T-cell epitopes in food allergens. However, two food allergens whose T-cell epitopes have been studied—Gal d2 and Ara h 1—have not demonstrated any unique properties.

On the basis of inducing nonresponsiveness and down-regulation of established immune responses in animals, there is considerable interest in peptides containing T-cell epitopes as a basis for immunotherapy in many allergies. However, practical application of immunotherapy with T-cell reactive peptides is still under study. B-cell epitopes do not appear to have any unique or common pattern of amino acid sequence. There are methods, although not totally successful, to predict B-cell epitopes. For example, algorithms based on the calculations of polar and nonpolar amino acid residues have been used to predict those residues located on the surface of the molecule. However, this information could be incomplete because host responsiveness may determine those epitopes to which an individual reacts. When the biological activities of various allergenic proteins have been compared, no consistent pattern representative for allergens in general or food allergens specifically has been evident.

Comparisons of primary amino acid sequences of allergenic proteins or tertiary protein structure has not yielded any unique or typical pattern. When the primary structure of an allergen is compared with other proteins, however, amino acid sequence similarities occur with many proteins in our environment. When considered in the light of the evolution of all living organisms, this finding should not be surprising but rather serves to illustrate that there may be certain as yet undetected structural features of allergens that render them different from other proteins.

## Implications for Biotechnology

As has been discussed in the previous chapters, recombinant DNA technology can have a profound effect in improving our food supply. Through these methods, crops can be improved by increasing stress, disease, and insect resistance and improving qualities desired by the consumers, including increased shelf life, better taste and flavor, and nutrition. Indeed, a variety of crops have been developed and approved by the Food and Drug Administration and are now available for use by consumers. With regard to allergenicity, recombinant DNA technology can have both beneficial and potentially detrimental effects. The ability to alter the structure of proteins or suppress their expression has a potentially beneficial effect by reducing allergenicity for foods with known allergens.

Conversely, expression of new proteins may result in new allergens or may quantitatively affect expression of homologous proteins that could be allergenic. Any of these possibilities may result in significant food allergy, and thus the allergenic activity of transgenic crops is an important concerns. Certainly it is known that recombinant proteins can be allergenic. Many recombinant allergens have been expressed and have been shown to retain their allergenic activity. This suggests that any recombinant protein allergen expressed in a transgenic crop has the potential to sensitize consumers who ingest this food. The implication for biotechnology when new food products are developed is that adequate testing for allergenic activity must be done before the release of transgenic food to consumer. When one considers the effects of biotechnology on altered foods, the following three issues are important:

1. Expression of proteins from known allergen sources and how to detect these allergens.
2. Expression of proteins form source of unknown allergenicity and how the potential allergenicity of these new products can be assessed. This is the most difficult effect to monitor because the potential allergenic effects are unknown.
3. Potential beneficial effect of biotechnology on food products through the development of hypoallergenic products.

## Measurement of Known Allergens by *In vitro* Assays (IgE Binding)

The expression of recombinant proteins with known allergic activity in different foods is the least difficult issue to address. To date, most experience with recombinant allergens has been obtained from recombinant inhalant allergens, and for the most part these recombinant allergens bind IgE antibodies and thus retain their IgE-reactive epitopes. These findings suggest that recombinant proteins expressed in other species (i.e. altered food crops) will probably retain their allergenic activity and must be considered allergenic unless proven otherwise. Can such allergenic recombinant proteins be detected in new foods?

Recombinant allergens in genetically engineered or altered foods can be identified using traditional in vitro assay such as the radioallergosorbent test (RAST) inhibition or the enzyme-linked immunosorbent assay (ELISA) inhibition assays. Both assays are based on the competitive binding of IgE antibodies in patient serum between the test sample in a liquid phase and the solid-phase allergen. Testing increasing inhibitor concentrations can be used to measure unknown allergen quantities in assay samples. These methods are well established, specific, sensitive, and reproducible. The only reagent of limited quantity is patients' sera. However, once sensitized individuals are identified, substantial quantities of sera can be obtained and serum pools produced.

At least 1 L of plasma can be obtained from sensitized individuals by plasmapheresis, which can provide sufficient reagent for virtually unlimited allergen assays. An alternate possibility would be to use monoclonal antibodies raised against such allergens. This would provide a potentially unlimited quantity of the reagent. However, monoclonal antibodies may not bind to IgE reactive epitopes; thus, the only definitive way to identify immunologically active allergens is through demonstration of an IgE reactive epitope.

**Examples of Allergen Detection in Transgenic Foods**

The importance of testing recombinant proteins for allergenicity has been clearly demonstrated. For example, the methionine-rich 2S storage protein from the Brazil nut was expressed in soybean to improve the nutritional quality of soy meal as animal feed. Following genetic transformation, the 2S protein from the Brazil nut constituted a significant fraction of the soybean protein. Because Brazil nuts are allergenic, in vitro tests, including RAST and immunoblots, were used to test extracts of the transgenic soybeans for expression of Brazil nut allergens. Using sera from Brazil-nut-sensitive subjects, 8/9 reacted to transgenic soybean extract, which confirms that an immunologically functional Brazil nut allergen had been transferred to the soybeans. Although Brazil nut allergy is not common, if the allergens were widely dispersed in a commodity food, such as soybeans, exposure, and presumably reactivity, would increase. Using RAST inhibition and immunoblotting, we assessed the allergenic potential of transgenic soybeans engineered to produce increased levels of oleic acid.

Virtually identical inhibition of wild-type extract and an extract prepared from transgenic soybeans was demonstrated. Results of immunoblotting demonstrated that both the wild-type and transgenic soybeans contained approximately 30 protein bands

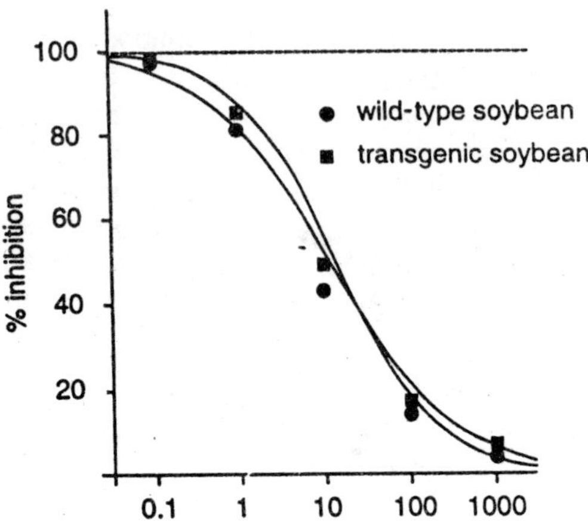


*Fig. 5.2. Inhibition of wild-type soybean RAST with wild-type or transgenic soybean extracts.*

ranging in molecular weight from 14 to 100 kDa; intensity of the bands appeared to be identical. The findings suggest that change in soy proteins resulting in increased oleic acid content did not alter soy allergen levels and, thus, do not pose increased risk of allergy to consumers. Lastly, we have investigated the allertenic activity of sulfur-rich corn proteins.

Two zein proteins, 10 kDa and HSZ, have been identified; because efforts are under way to increase the expression of these proteins in corn or the seeds of other cereal grains to enhance the sulfur content of sulfur-poor crops, it is important to assess their potential allergenicity. Sera from 23 individuals demonstrated by skin test, RAST, or clinical history and immunoblot to be corn reactive were tested for IgE antibody reactivity to these two zein corn proteins using SDS-PAGE/ immunoblotting. None of the sera from corn-reactive subjects demonstrated IgE reactivity against either the 10 kDa or HSZ zein proteins, which suggests that products encoding these genes do not pose an increased risk or allergy to consumers.

## Assessing Allergenicity in Genetically Engineered Food Derived from Sources of Unknown Allergenicity

A much more difficult problem than the one outlined above is evaluation of the allergenic potential of foods engineered using proteins from sources of undetermined allergenicity. Predicting potential allergenicity is a major challenge for the food industry because there is no single predictive assay to assess the potential allergenicity of any protein. A preliminary step, evaluation of amino acid sequence homology, may be useful in evaluating allergenicity of a transgenic proteins. If sequence homologies are observed, particularly with regions containing IgE binding epitopes, it is essential that in vitro testing with RAST or ELISA should be performed to assess IgE reactivity.

If no IgE antibody reactivity is detected, the second step should be compare the physicochemical and biologic characteristics—including molecular size, stability, solubility, and isoelectric point—of these proteins with major food allergens. Suspect proteins should again be tested to assess IgE reactivity. Clearly, these approaches may be extremely difficult, if not impossible, to establish if IgE reactivity is not known. Even the most careful evaluation may not exclude a potential allergenic risk, and it may be prudent to follow new food after their introduction into the marketplace.

**Screening Amino Acid Sequence Homologies with Known Allergens**

Because allergens from food and nonfood sources can cross-react, it is critical to compare amino acid sequences from all known allergens (not just with food allergens) with those from the genetically engineered food. Over 200 allergens have been identified, characterized, and sequences; this information is available through public domain databases (i.e. GenBank, PIR, EMBK, and Swiss Prot). On the basis of the optimal peptide length for binding B-cell epitopes (8-12 amino acids), an immunologically significant sequence would require a match of at least eight contiguous identical or similar amino acids. Confirmation of such a sequence suggests that the transferred protein may be allergenic and in vitro testing, as described above, should be employed to assess further the allergenicity of the protein.

Failure to find such a match strongly suggest that the introduced protein does not share linear epitopes with known allergens but does not exclude the possibility that the transferred protein is an allergen. Linear epitopes are important; however epitopes that are dependent on tertiary structure of the protein (conformational and epitopes) are not as easily assessed and are not readily predicted based on the primary amino acid sequence of the allergen. Moreover, amino acid sequence similarity does not necessarily mean similar IgE reactivity, for the degree of homology must also be considered. Recently it has been shown that a single amino acid substitution in the major peanut allergen Ara h 2 could abate, diminish, or enhance IgE binding depending on the substitution. Thus, for a 10-residue region of a transgenic protein, an amino acid sequence identity of 90% with a known allergenic epitope may indicate possible allergenicity but does not prove clinically significant cross-reactivity.

## Comparison of Physicochemical and Biological Characteristics

As stated above, food allergens generally share physicochemical and biologic characteristics, including molecular size, stability, and isoelectric point, thus, it is beneficial to compare these aspects in proteins of transgenic food with known food allergens. Of these traits, stability to digestive processes may be an important factor when assessing potential allergenicity. Proteins that are stable under proteolytic and acidic conditions of the digestive tract are more likely to reach the intestinal mucosa and stimulate an immune response than more labile proteins. Thus, rapid degradation of proteins expressed in genetically engineered foods reduces the likelihood that the protein is an allergen. Recently, an in vitro model to assess the stability of food allergens to digestion was developed. In this model, known food allergens and other common plant proteins were exposed to simulated gastric fluid for varying periods of time.

Food allergens were stable under these conditions for periods up to 2 min, and some major allergens were stable for more than 1 h; whereas, the few nonallergenic food proteins tested were degraded within 30s. It should be emphasized that stability of a transgenic protein is simulated gastric acid is not synonymous with allergenicity because some allergens are rapidly degraded by proteolytic enzymes and many non-allergenic proteins can be very stable to proteolysis. Thus, there needs to be more detailed studies on the enzyme stability of more allergenic and nonallergenic food proteins in order for enzyme digestability to considered as a major property of allergens. Lastly, the prevalence of the protein in a food should also be considered.

In plant foods, many food allergens are storage proteins that are present in large amounts; thus, many food allergens constitute a large proportion of total protein (1 to 80%) in offending foods.

Introduced proteins are usually expressed in plant in very low levels; from less than 0.01 to 0.4%. It is tempting to speculate that degree of exposure to allergens (i.e., the amount of an allergen in a food) is directly related to allergenic potential. However, the major allergen is codfish, Gad c 1, is not a predominant protein, and proteins that are major components of many foods, including myosin, tropomyosin, and actin from beef, pork, and chicken have not been identified as major allergens.

### Biotechnologically Altered Foods

There can be a beneficial effect on allergenicity of transgenic foods eliciting allergenic responses can be rendered hypoallergenic through deletion of the allergen or modification of the allergenic epitopes. This technique is potentially very useful because it can substantially reduce a food's allergenic activity. Recently, genetic engineering of hypoallergenic rice has been accomplished. A rice seed protein of approximately 16 kDa was isolated and identified as a major rice allergen. A cDNA clone encoding this protein was isolated from a cDNA library prepared from maturing rice seeds.

On the basis of the nucleotide sequence of the cDNA, an antisense RNA strategy was applied to repress expression of this allergen in maturing rice seeds. Seeds from transgenic plants with the antisense gene have substantially reduce amounts of the allergen relative to the wild-type control. Recently Burks and colleagues have investigated two major peanut allergens: Ara h 1 and Ara h 2. Four major IgE binding epitopes have been identified in Ara h 1, and three major IgE binding epitopes were identified in Ara h 2. Through a single amino acid substitution, IgE binding to peptides containing three of the four major epitopes of Ara h 1 and all three epitopes in Ara h 2 was abolished. Current studies are directed at investigating the effects of such amino-acid substitutions on the entire allergen molecule. Together, these two studies are very exciting because they demonstrate that biotechnology can be used beneficially to suppress or alter expression of major allergens.

## Conclusions

Recombinant techniques offer breeders, horticulturists, farmers, and consumers agricultural crops with improved qualities, including increased resistance to stress, insects, diseases, and herbicides. Because genes governing the coding of proteins conferring these new traits are not part of the plant's original genome, there are concerns for the safety of these newly engineered varieties. A major concern is the allergenic potential of transferred proteins. Systematic strategies to assess potential allergenicity of these proteins have been proposed; these strategies will be modified as more is known about the structural basis of allergens. The main problem is evaluating the allergenic potential of proteins from sources that have not been implicated as allergens.

Current approaches include comparison of amino acid sequences of the transferred protein with those of known food and nonfood allergens and comparison of chemical and physical characteristics of the transferred protein with known food allergens. In vitro assays using sera from hypersensitive individuals or placebo-controlled double-blind food challenges should be used to confirm allergenicity of any transgenic foods before marketing. Although this chapter has focused on problems arising form transferring known allergens or creating new allergens, bioengineering techniques also provide a unique opportunity to create hypoallergenic varieties of plants. Matsuda and coworkers genetically engineered a hypoallergenic rice.

By introducing genes in the antisense orientation, the levels of expression of the 16-kDa rice allergen were reduced when compared with the wild type. The exciting study by Stanley and associates suggests that altering IgE binding epitopes of the peanut allergen Ara h 2 could substantially reduce allergenicity of this ubiquitous food. Thus, biotechnology may be used to alter the allergenic potential of foods.

# Chapter 6

# PRODUCT DEVELOPMENT

During the brief history of modern cereal biotechnology, the development of innovative products has captured a great deal of academic interest and corporate attention, with such innovation being viewed as a highly visible source of competitive advantage. However, while there has been extensive basic research on the problems of characterizing genes with specific agronomic effects, very little has been written about the problems associated with product commercialization in cereal biotechnology. The increasing complexity of transgenic plant products, especially those containing metabolically engineered output traits, will raise the cost of launching the next generation of products in cereal biotechnology will have to pursue two seemingly incompatible goals: increasing product differentiation while reducing manufacturing costs and development times in an effort to commercialize effectively their products.

## Product Development versus Process Development in Cereal Biotechnology

Product development focuses primarily on designing and testing prototypes of the product. Process development can be described as a system for creating and refining an organization's capability to manufacture a product or set of products commercially. While there is a large body of literature on the competitive advantage of efficient process development in mature industries, such as bulk chemicals, and an increasing focus on 'design for manufacturability' in assembly products such as automobile production, little has been published on the competitive role of effective process development in agricultural biotechnology.

Within cereal biotechnology, decreasing process development lead times should result in accelerated product development lead times, especially where innovations in process design are implemented on the road towards transgenic product launch. Rapid and effective ramp-up is essential for lowering the need for capital investments in breeding capacity, lowering current production costs, generating sales revenue and recovering development costs. Some of the resulting reduction in manufacturing costs can be passed on to farmers and other consumers to increase end-user acceptance of new products. Finally, innovative process design technologies can extend the proprietary position of a specific cereal biotechnology if would-be imitators or competitors are unable to determine how to produce the transgenic product either at competitive costs or at sufficient quality levels.

In this chapter I would like to describe how innovative process development, leading to rapid product commercialization, can generate a competitive edge in cereal biotechnology. A review of the overall product R & D cycle for developing transgenic cereals, using transgenic corn commercialization in the USA as an example should help illuminate how and where various process development activities fit into the broader product development efforts within cereal biotechnology.

The product R & D cycle for transgenic plants can be divided into a number of stages that can include: discovery, small-scale greenhouse efficacy screening and evaluation, the generation of intellectual property protection and regulatory approval, large-scale field evaluation and breeding and product release. Finally, towards the end of the chapter, I will describe the development of transgenic corn products expressing B.t. transgenes. This example should illustrate how the contribution of each step in process development cycle leads to the final commercialization of this particular product.

## Commercial Targets for Cereal Biotechnology

One of the major goals of discovery research in cereal biotechnology is to identify genes that safely and effectively generate commercial opportunities in agriculture. The challenge for product designers in cereal biotechnology firms is to close the gap between what product can be offered to the customer, i.e. technology push, and what the customer really wants, i.e. market pull. Technology push is a result of novel product opportunities arising from, for

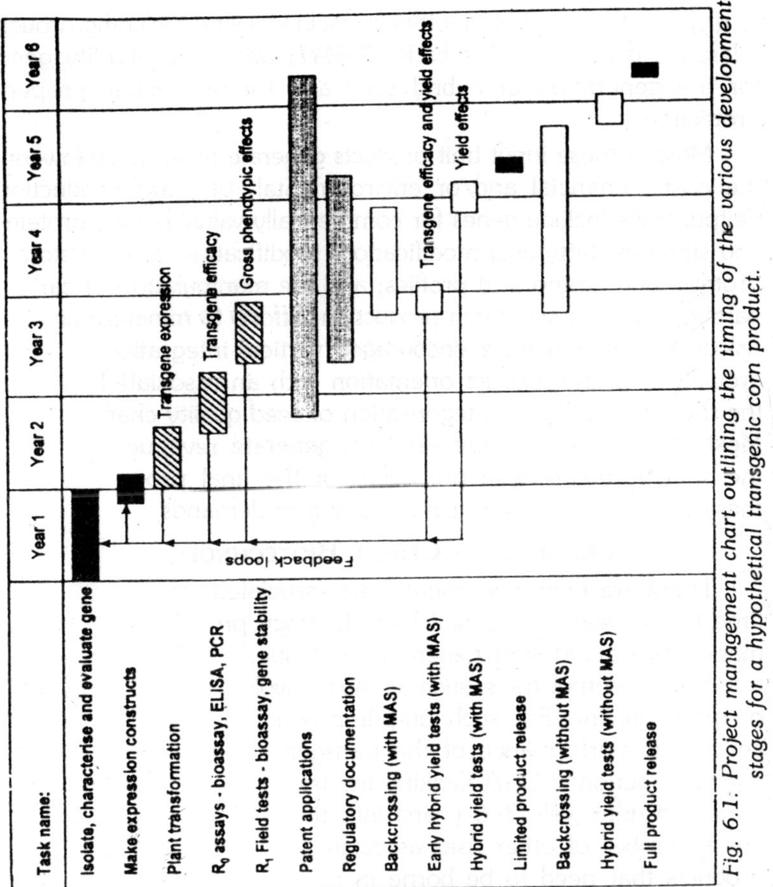


*Fig. 6.1. Project management chart outlining the timing of the various development stages for a hypothetical transgenic corn product.*

example, advances in genetic engineering, genomics, information management systems and grain characterization technologies, while market pull comes from increasingly sophisticated end-user and consumer demand for improved processes and products.

The product traits themselves can be broadly divided between a first generation of relatively simple input traits, some of which are already on the market, and a second generation of increasingly complex output traits, some of which are about to come onto the market or are in the product pipeline. Input traits include genes for pest and disease tolerance, such as B.t. genes α-amylase inhibitors, viral coat proteins, and viral replicase genes; herbicide resistance, including those derived from mutation screening, e.g. resistance to sethoxydim and transgenic approaches, e.g. resistance to glyphosate and phophinothricin; yield stability, including drought tolerance genes, such as barley *HVA1*; and male sterility genes for the generation of hybrid seed and for intellectual property protection.

Most of these input trait products generate revenue by lowering the costs, financial and/or environmental, of plant production. Output traits include genes for commercially valuable oils, proteins, and starches; fatty acid modification; modification of seed storage proteins and amino acid profiles; and the manipulation of carbon-partitioning for novel starch production. Efforts to minimize financial risk in agricultural have encouraged vertical integration, contract growing and end-product orientation with an associated focus on the measurement and categorization of seed quality characteristics. Most of these output trait products generate revenue by altering the specific measurable properties of the final product to meet these increasingly sophisticated consumer demands.

## Problems in Cereal Biotechnology

There are high investment costs associated with the long lead time (6–10 years) for cereal biotechnology products to reach the market place and start generating revenue, and this is especially true for the output traits, such as those involved in the modification of grain quality. For such output products the impact on final agronomic performance of the transgene, and/or the effects of the plant genomic DNA flanking the transgene, might not be fully evaluated until well (3–6 years) into the breeding process. There are a number of other risks associated with cereal biotechnology products that need to be borne in mind during the initial design phase. There can be uncertain profitability associated with some of the 'technology push' concepts at the onset of product development. There may be intellectual property tissue limiting a company's freedom to operate with key technologies or encouraging the rapid development of potentially superior technologies by their competitors.

Finally, there might be uncertainty associated with regulatory and consumer acceptances issues inhibiting trade and domestic investment, for example, the export of processed grain derived from transgenic plants to the EU. New products coming out of discovery research do not necessarily ensure lasting profits. Returns form innovation can be competed away unless isolating mechanisms are in place to inhibit imitators. As the mean development time to launch the next generation to output products increases, the time lag between gaining regulatory approval and patent expiration will inevitably shrink. Although patent protection lasts a number of years following the date of initial filing, some portion of this period will be lost because of the time required to 'fine tune' the technology, conduct efficacy trials, secure regulatory approval and produce the finished transgenic product in sufficient sales volumes. Instead of conducting their own lengthy and costly regulatory procedures, the products of generic (off-patent) cereal biotechnology products, which could start coming onto the market in the next decade, might need only to show that their version of the product is chemically an biologically equivalent to the original patented version. Therefore, for off-patent products, firm-specific differences in their process knowledge base will become an increasingly valuable source of competitive advantage.

## Efficacy Screening of Commercial Traits

The development of cereal biotechnology process plan begins with the design of those analytical methods, screens and other assays that will be used to detect the expression of the desired gene in early generation plants. However, the evolution of analytical methods should continue throughout the project, with later development emphasizing methods applicable to production quality control during breeding and foundation seed production.

Analytical techniques play a critical role in evaluating process R & D experiments throughout the product development life cycle by helping to generate a deeper understanding about underlying cause-and-effect relationships in the recombinant trait's efficacy. Particularly novel products will require the invention of novel analytical techniques, tools and procedures. The ability to recognize the need to develop novel analytical techniques in a timely manner is an important capability within any product commercialization group. Screening methods generally fall into two major types, selective screens or analytical screens.

The development of selective screening systems, e.g. for herbicide tolerance, may begin in tissue culture, followed by initial greenhouse trials and experimental fields where promising transgenic lines are evaluated in randomized, small-scale plots. The development of such selective screening systems has been important in the evaluation of input trait efficacy, where many of the products can be evaluated in early generations in genotypes

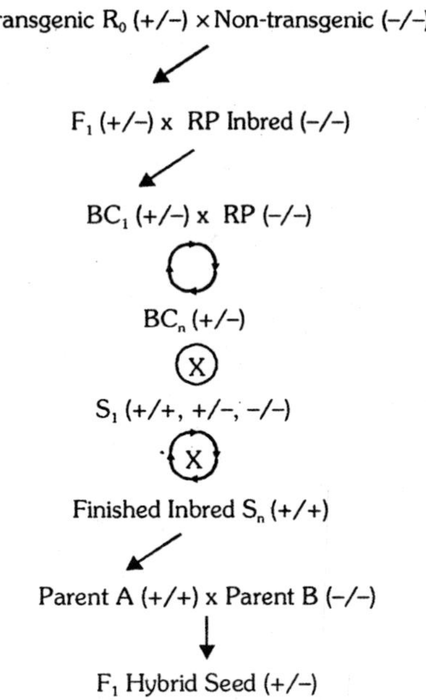


*Fig. 6.2. Outline of the breeding process for a transgenic corn product. The genotypes of the various generations are indicated as either present (+) or absent (–) for the transgenic trait.*

that are readily transformed but are generally not commercial varieties. Analytical screening systems, e.g. for transgene presence, event verification, oil content, amino acid composition and other grain quality characteristics, may be followed by a subsequent functional assay if such an assessment is critical for the initial identification of efficacious lines. For those output traits affecting grain quality characteristics this may involve small-scale compositional analysis followed by large-scale animal feeding trials or large-scale processing experiments.

Many grain quality characteristics can be fully evaluated only in the finished inbred line either to be sold directly or used in hybrid seed production. The generation of such finished lines is both time consuming and expensive. During the early generation evaluation phase of the product development cycle, the company's goal is to collect sufficient data on the performance of the candidate gene to warrant the much more expensive step of large-scale field evaluation and entry into the breeding program. At this early stage gene candidates can be abandoned and/or targeted for redesign as a result of either a lack of demonstrated efficacy, gross agronomic abnormalities or phytotoxicity. The precise principles governing the expression of recombinant traits in transgenic plants are poorly understood. As a result, problem solving in cereal biotechnology relies more heavily on physical experimentation than on conceptual modelling. Thus, while early greenhouse observation are critical, the real uncertainty lies in predicting a recombinant trait's agronomic performance in elite commercial germplasm. This is especially true for output traits the efficacy of which can be significantly influenced by genotype effects. In this environments on tends to find that 'learning by doing' is the only available process development strategy.

## Molecular Breeding of Transgenic Plants

The aim of the breeding program is to develop high-yielding competitive varieties free of any yield drag or other agronomic problems that might be associated with either the novel gene's expression or its initial genetic background. This breeding process might take 5–9 years to complete in corn, depending upon the number of backcrosses involved. Once early ($R_0/R_1$) generation transgenic lines have been characterized they will entered into a breeding program with a number of quality inbred lines to do initial assessments of genotype and environmental effects on the performance and genetic stability of the new trait (1–2 years).

The efficacy of the novel trait must be assessed through several seasons in multiple locations to ensure that performance standards are consistently met in a number of unique environments. Those transgenes and transgenic events that make it through this initial field evaluation will eventually enter into a variety development program, involving crossing into other high-quality lines (3–5 years). Towards the end of varietal development, seed production begins to generate enough seed stocks for commercial release through normal foundation seed multiplication channels (1–2 years). Individual transgenic 'events' are eliminated at each step in the breeding program as they fail to meet performance benchmarks, so the initial evaluation process must start with a large number of high-quality transgenic lines.

### Shortening Breeding Programs

A number of processes can be used to shorten transgenic breeding programs. The development of transformation protocols for cultivars containing a genetic background common to that of elite commercial varieties will reduce the number of backcross generations needed to produce a fully converted transgenic inbred line. Winter nursery facilities can be used for year-long production of multiple product generations. Double-haploid plant production can be used to produce homozygous transgenic lines for rapid evaluation of transgene copy number effects, yield drag associated with individual insertion events or combining ability for stacked products containing more than one transgene. Finally, marker-aided selection can be employed to facilitate greatly the selection of progeny for use in subsequent backcrossing to recurrent parent lines.

### Marker-assisted Selection for Transgenic Plants

Valuable alleles and/or transgenes can be tracked in breeding populations using genetically linked molecular markers. A number of marker systems are available for this process including: restriction fragment length polymorphisms (RFLPs); simple sequence repeats (SSRs); amplified fragment length polymorphisms (AFLPs); and single nucleotide polymorphisms (SNPs). Marker-aided selection (MAS) is used to select for genetic similarity to the targeted elite inbred (recurrent parent) within early backcross generations. MAS facilitates the elimination of a number of generations from the backcross conversion process and might save 1-2 years in transgenic product development. MAS also allows one to select for less linkage drag associated with the 'non-elite' genomic DNA flanking the transgene, increasing the probability of obtaining an acceptable conversion.

It is advisable to use a sufficient number of markers to sample the genome adequately, with the actual number depending upon the genetic relationship of the transgenic line to the recurrent parent, the backcrossing strategy and the breeder's experience. Uniformly distributed markers are more effective in backcross conversion and minimizing linkage drag than an equal number of markers per chromosome or randomly assigned markers. Using MAS to accelerate backcross conversion is critical to the introduction of value-added traits in elite genetic backgrounds, the efficient management of seed inventories and the timely commercial launch of new transgenic products.

## Molecular Quality Control for Transgenic Plants

Transgenic plant lines containing independent transgenes and independent transgenic insertion events are backcrossed and evaluated at a number of different breeding locations. This is done,

(a) Molecular breeding: $BC_1$ line production

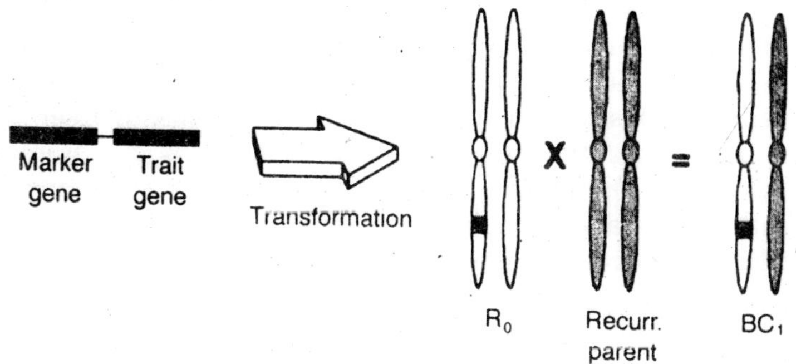


(b) Molecular breeding: transgenic hybrid production

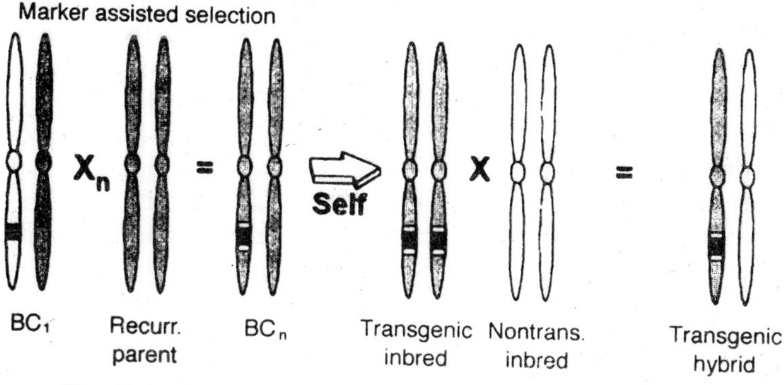


Genetic marker spread = 1 every 15cM, 1cM = 1.2 Mbp in corn
non-elite gDNA ~ 7.5 cM ~ 9 Mbp

*Fig. 6.3. Overview of breeding a transgenic corn product.*

in part, to evaluate transgene efficacy under a number of different environmental conditions. Within this breeding and evaluation process there exists the potential for line and/or gene misidentification. It is important that events should not be mixed up in breeding process since only one event might be deregulated for eventual sale. Finally, at the end of the breeding process, the customer must be provided with an estimate of the quality of the transgenic product that they are purchasing, where quality is represented by the percentage of 'off types' they can expect in their field of transgenic plants.

In breeding programs transgenic lines are often identified by screening for the presence of a more readily detectable selectable marker gene (e.g. herbicide resistance) that was co-introduced with the transgene of interest. Insertion event-specific transgene silencing, recombination-mediated transgene deletion and/or pollen 'contamination' may result in loss of expression for the non-selected transgene of interest. Without a quality control system in place, generation of selection for the readily detectable marker gene may pass before the presence/efficacy of the transgene of interest is evaluated. Therefore, independent transgenes and transgenic events must tracked and verified throughout the breeding and evaluation process to ensure that transgenes are not 'lost' or transgenic lines misidentified. This gene tracking process has been dependent upon the development of a number of analytical tools, including rapid transgene and insertion event-specific assays.

The gene tracking process has also been facilitated by developments in the automation of DNA isolation/analysis and in data processing. The recent evolution of quality control systems in cereal biotechnology is a good example of prior production experience, or 'learning by doing', generating data that can contribute to future process development projects, thus allowing for 'learning' to take place 'before doing' subsequent process development projects. In relatively immature industries, such as cereal biotechnology, manufacturing performance should improve with cumulative experience. Cumulative production experience generates data for systematic problem identification and problem solving, leading to improvements in product design, equipment modifications and worker training. However, for such 'learning by doing' to be retained and utilized, there need to be strong feedback loops from the production process (in this case the breeders) to process R & D teams.

## Intellectual Property and Freedom to Operate

Freedom to operate (FTO) under any dominating patents must be secured in order to commercialize a product. Transgenic plants often utilize several patented technologies, each owned by different parties, e.g. transformation method, selectable marker gene, specific genes, and gene expression elements. Therefore, companies must commit substantial resources to develop either in-house, or acquire rights to utilize, all of the technologies used in the development of their commercial products. These issue should ideally be addressed before the product enters the breeding program for large-scale field evaluation. In order to obtain a US patent, the technology must be novel (no prior, art) non-obvious (no analogous prior art) to one ordinarily 'skilled in the art', and have some defined utility.

The conception and reduction to practice of claimed invention (the actual invention and any modifications to the invention that retain the same activity that are taught in the patent application) should be described in such a way that anyone in the field could make and use the invention. From a product development point of view in cereal biotechnology patents may be of essentially two major types. A product patent covers compositions of matter such as genes and gene products, monoclonal antibodies, engineered cells, and transgenic plants.

A process patent covers methods such as those for gene cloning, transformation, gene silencing, gene delivery, and molecular breeding. There are a number of problems associated with patents with respect to process development. First, there is usually a long delay between an initial patent application and the issuing of the patent. In the US, the contents of a patent applications are not released until the patent is actually issued. Second, there are often a number of overlapping claims between related patents and there is often inconsistency in the breadth of coverage granted between different countries. Third, there can be significant resource requirements to file the patents and, especially in cereal biotechnology, to enforce the patent. Finally, overly broad patent protection is normally not granted at the present time.

Such broad patents are generally not written in such a way as actually to teach anyone skilled in the art how to repeat all of the application's claims and the granting of such broad patent protection is also believed to inhibit subsequent innovation. All of these patent issues, while problematic from the biotechnology firm's point of view, have been a boon to patent litigation lawyers. Thankfully there are a number of solutions to these patent problems that should be investigated during the product's development.

Licensing and cross-licensing of technology between firms can occur in order to exchange rights to patents critical to product commercialization. Mergers and acquisitions, either partial or complete, can be orchestrated in order to gain patent rights for critical technological component(s) in the commercial product. Finally, firms can engage in technology development to supersede a competitor's patents.

## Regulatory Issues and Risk Assessment

There is a widespread belief that standard plant-breeding procedures are generally not adequate to identify all the potential problems associated with transgenic plants. It is argued that, as a consequence of the diversity of available transgenes, the environmental consequences of genetic modifications cannot reasonably by anticipated. Therefore, countries have established their own agencies to oversee the introduction of new transgenic plant varieties into their environments and the importation of novel plant products, particularly transgenic, into their marketplaces.

In the establishment of these national agencies, most governments have striven to balance the need to develop oversight mechanisms that address potentially serious risks without generating a regulatory burden which creates disincentives to innovation and diminishes the availability of important new techniques and products. Securing government regulatory approval for product introduction and addressing customer acceptance issues are critical aspects of product development and commercialization, e.g. for key export markets. In addition to gaining approval of the transgenic 'event' itself e.g. for herbicide resistant crops, the registration and deregulation of any associated chemicals used with the transgenic varieties may also be required.

In most countries there are two kinds of regulation that govern research and development of transgenic plants. Rules for 'contained use' govern genetic modification in the laboratory, concentrating mainly on worker health and safety issues. In the US the rules for 'field release' focus on environment on environmental risk assessment appropriate to the nature and final use of the transgenic plant, with each release initially considered on a case-by-case basis in order to build up experience with particular crop/transgene combination.

A number of factors are borne in mind during the risk assessment process associated with transgenic plant deregulation. These factors include: the function of the gene in the donor organism; the effect of the transgene on the phenotype of the transgenic plant; the risk of the transgenic to animals and humans, e.g. evidence of toxicity and/or allergenicity; the ability of the transgenic to colonize and persist in agricultural habitats (weediness); and the proximity of the transgenic species to its 'centre of origin'. These last two factors address the likelihood and consequence of transgene movement to other weedy or wild relative plants by cross pollination or to other (pathogenic) organisms by horizontal gene transfer.

Other risk issues include the impact of the transgenic plant on the evolution of target organisms, e.g. insect resistance to B.t.; the behaviour of non-target organisms, including minor pests or beneficial insects; the existing ecological relationships within the agricultural system, such as the plant's potential to disrupt existing pollination systems; and the ability of the transgenic to provide a resource for organisms that are pests of other crops within the agricultural ecosystem. Individual nations generally employ one of two different kinds of regulatory systems, with the data assessed in both being similar. Horizontal regulatory systems, such as those within the EU, are process-based systems that apply to all plants produced by a certain method, e.g. by a particular transformation method. Vertical regulatory systems, such as that in the US, are product-based systems that define the characteristics of modified plants that require them to be regulated.

So, except for FDA approval, the US has a techniques-based regulatory system which differentiates between 'natural' exotics, genetic improvements by conventional methods, and manipulation by recombinant DNA techniques. Within the US, transgenic plants are regulated by three government agencies, following a 1986 decision to regulate transgenics under existing statutes. The US Department of Agriculture (USDA/APHIS) controls permits for inter-state movement and field release of transgenic materials, assesses the pest potential of transgenic plants and determines when transgenic can be field-grown without notification or permits, otherwise known as deregulation.

The food an Drug Administration (FDA) determines that a transgenic plants has been adequately evaluated in accordance with its biotechnology food and feed policy, e.g. for the safety of antibiotic selectable marker genes. Finally, the Environmental Protection Agency (EPA) regulates transgenic plants with pesticidal properties, e.g. B.t. plants, and determines pesticide/herbicide residue tolerance on transgenic plants, and determines pesticide/ herbicide residue tolerance on transgenic plants. In Canada, transgenic plants are regulated by Ag. & Agri-Foods Canada and Health Canada. In the Canadian system, products are similarly regulated whether generated by mutation or transformation. This is essentially a risk-based system that focuses on the intrinsic properties of the product, e.g. weediness, invasiveness, or its potential for outcrossing, rather than on the method of production of the variant under evaluation.

In Europe, a Novel Foods Regulation went into effect in 1997 to replace completely EU Directive 90/220 that previously covered environmental issues associated with transgenic seed sales and transgenic grain import. In Japan, transgenic plants are regulated by the Ministry of Agriculture, Forestry and Fisheries and the Ministry of Health and Welfare. To date, there is no international harmonization of regulations to ensure that transgenic plant varieties released in one country will be accepted in another, so antibiotic resistance genes in food products might inhibit international trade in transgenic products. Meanwhile, The Convention on Biodiversity has established a biosafety working group to develop safety standards for international trade in transgenic products.

## Product Release and Marketing Strategies

In order to recover its substantial R & D investments, the developer of a commercial product might either generate and market the transgenic seed directly or negotiate a royalty with a seed company/companies, e.g. for technologies discovered by universities, government agencies, technology development companies or large agrochemical companies. Because of the previously described learning curve effects in process development, new products might have higher manufacturing costs than older products. Thus the profit margins on a newly launched product might be small or even negative during the first few yeas of its commercial life. In transgenic plants, the product is carried in a vegetatively or sexually reproduced form which the customer could potentially propagate following its initial purchase.

For hybrid crops, new seed must be purchased each year; therefore the transgene is protected and premiums for the technology can be added to the seed. However, for inbred or vegetatively propagated crops, where transgenic material could be replanted each year, the transgene is not biologically protected; therefore farmers might be obliged to take a licence from the seed company that allows them to buy and grow seed containing the transgene. Alternatively, firms might need to develop biological processes to protect their investment, such as the so called 'terminator' technology. The marketing strategy for any one particular trait will be influenced by the nature of the product.

Input traits, such as those conferring herbicide, disease or stress tolerance, which enhance yield or reduce inputs, can return revenue either by increasing the seed company's market share and/or increasing seed sale premiums. For some products, such as herbicide tolerant crops, the agrochemical companies that helped develop these products can benefit from increased chemical sales, e.g. Roundup-Ready® Corn or Liberty-Link® Corn. For output products, such as improved grain quality, there may be either a seed sale premium, an end-product premium for contract-grown crops or a mechanism for sharing the added value with the end-user, such as a food-processing company.

## Product Development: Practical Example

A review of the overall product R & D cycle for developing transgenic B.t. corn, should help illuminate how and where the previously described product and process development activities fit into the broader product commercialization efforts within cereal biotechnology.

### B.t. Genes, Proteins and Mode of Action

*Bacillus thuringiensis* bacteria produce insecticidal proteins, each with their own spectrum of activity. These accumulate as crystals and are thought to aid in bacterial spore germination and vegetative growth following ingestion and gut wall destruction in certain vector organisms. B.t. protein specificity is determined by its interaction with aminopeptidase-N receptor molecules on the luminal side of the insect midgut epithelium. B.t. protein toxicity appears to be determined by the formation of a pore in the cell membrane, which leads to cell lysis and disintegration of the midgut epithelium. The B.t. crystal protein consists of three domains: a helices responsible for pore formation; a receptor binding domain; and a third domain of unknown function. By exchanging corresponding domains between crystal proteins their insecticidal activity can be exchanged, expanded and /or improved.

There are a number of potential advantages associated with the development of transgenic B.t. crops. Their development and deployment will reduce the cost, time and effort spent protecting crops from insects and, relative to many commercially available chemical insecticides, they should contribute to an environmentally friendly crop production system.

## Development of Transgenic Crops Expressing B.t. Transgenes

B.t. corn with improved European corn borer (ECB) resistance was first released to farmers in 1996 and a number of competing B.t. corn products are now on the market. However, a number of product design and process development issues had to be addressed in order to develop and commercialize transgenic crops expressing B.t. transgenes. Efficient transformation and selection system had to be developed for all major crops with the choice of transformation protocol used by any one company being influenced by intellectual property considerations.

Efficient transgene expression systems needed to be generated. These included the use of synthetic B.t. genes optimized for plant codon preference, truncated B.t. genes, and highly active constitutive promoters for driving high level of expression in transgenic plants. Companies had to establish efficient transgenic plant production systems. Gene transfer can result in multiple transgene integrations and rearrangements of the heterologous and/or homologous DNA with associated effects on transgene expression. In many cases a large number of transgenic events needed to be produced in order to identify the few lines with adequate transgene expression levels and agronomic performance.

Transgenic plant identification and characterization methods had to be established and validated. Molecular breeding and molecular quality systems had to be put in place to handle transgenic products. New regulatory systems for transgenic plants had to established. For example, in the US the EPA considers B.t. crops as pesticides and they need to be registered as such, including toxicological studies on a range of organisms. Companies hoping to sell B.t. products needed to design and implement resistance management (RM) strategies, with grower recommendations and monitoring protocols to ensure that resistant insects did not evolve so that the lifetime of the product could be maintained. Systems had to be developed to integrate these new technologies into current farming practices. Finally, there was, and continues to be, a need to establish acceptance of transgenic crops and crop products by growers, food processors and consumers.

## Insect Resistance Management

The success of B.t. crops will depend on whether target pests develop resistance to them. Specific inset resistance management

(IRM) measures have been, and are continually being, designed and implemented to minimize the build-up of resistance genes in insect populations that might develop under constant selection pressure from transgenic crops. A number of different IRM strategies have been proposed. In the 'gene stacking strategy' transgenic plants are produced that express a combination of different B.t. genes, or other insect resistance genes, each with a unique mode of action that should delay the evolution of resistant insect populations. In the 'high dose + refuge strategy' transgenic plants are developed with a *high dose*, or relatively high level of B.t. protein accumulation, to eliminate resistance genes in heterozygous resistant insects that are only slightly less susceptible to B.t. than fully susceptible insects. At the same time a *refuge* of non-B.t. plants is provided to dilute resistance development in insect populations. In 1998 Monsanto's IRM Strategy for their YieldGard® B.t. corn allowed growers either to plant at least 5% of their acres with a non-B.t. corn hybrid and not treat the rest with insecticides registered to control ECB or to plant at least 20% of their acres with a non-B.t. corn hybrid and treat all their corn aces with non-B.t. insecticides as needed.

**Commercial Goals of Insect Resistant Corn Products**

The goal of insect resistant corn is to redirect a substantial portion of the insecticide market from agrochemicals to the seed industry. Almost 25% of all pesticides used in US agriculture are insecticides. In 1996 the insecticide market was estimate to be ~$7 billion. However, for transgenic plants expressing B.t. genes, it should be borne in mind that their initial ECB target is a cryptic feeder that spends most of its time in the stalk and so it is a pest that is inaccessible to most conventional insecticides. Therefore, it is uncertain exactly how much of the conventional insecticide market has been captured by B.t. plant products. The B.t. technology itself is an insurance product, with the return of the farmer being directly proportional to the level of potential yield loss that would have been caused by insect damage in the absence of any protection afforded by the B.t. transgene. The break-even point is estimated to fall between 2-4% potential yield loss, depending upon the yield potential of the crop.

During 1997 the actual yield advantage for B.t. corn over non-B.t. corn in the mid-west US was determined to be between $15 and $43 bushels/acre. From the onset B.t corn was expected to increase yield by ~5% in ~20% of the 82 million acres of hybrid corn planted in 1998, representing a potential gross added value of ~$350 million to those who developed this transgenic cereal product. B.t. corn sales in 1998 represent an added gross premium of ~$150 million, with the final net added value being dependent upon the costs associated with seed production, quality control, royalties to third-party technology providers, market share gain, discounting, dealer costs, etc.

## Future Trends

Efforts to minimize financial risk will continue to encourage functional consolidation, vertical integration, contract growing and end-product orientation in modern commodity agriculture. The resulting agribusiness conglomerates should have a number of common enabling features in order for them to take full advantage of future product commercialization opportunities in cereal biotechnology. They should have access to proprietary, elite germplasm so that they can piggy-back their traits onto already high-yielding varieties. They should have the ability rapidly to move traits into finished products with an extensive, efficient seed presence (sales > 150,00 units/yr) in order to capitalise fully on the investment, direct or indirect, required to obtain the transgenic trait. Finally, hey need to have innovative customers who are capable of responding to new product opportunities. From the description of the product R & D cycle in cereal biotechnology, one can also conclude that a major challenge facing firms today is to create development processes that are fast, efficient, and capable of generating high-quality products.

While process development can have a significant impact on a product's manufacturing cost, savings can be missed by failure to invest adequate resources in process development early enough in the product life cycle. It should now be apparent from the foregoing description that development projects have two major outputs, the technology that is implemented in the new process and organizational knowledge that becomes available for future projects.

Organization knowledge includes how to manage projects, allocate resources, assign personnel and resolve disputes. Each new project should generate feedback on gaps between expected and desired lead times, uncover critical parameters, and create insights about the process performance that triggers a search for new approaches to managing future development projects. Much of this new knowledge becomes embedded in the firm's organizational routines and procedures. However, high performance in process development requires the ability to learn, and some organizations will be better able to learn from their experiences than others. Learning is rooted is how organizations feed data from manufacturing (breeding, foundation seed and sales) back into its technical knowledge base allowing R & D to anticipate future problems.

Organizational or geographical barriers between process R & D scientists are incapable of designing processes that work well in the field. In conclusion, the success of any individual organization will depend upon its ability to do more proficiently than their competitors those things that their products demand and their customers value. As cereal biotechnology matures, and consumer prices eventually stabilize, the ability to develop rapid and efficient processes will be an increasingly important source of competitive advantage in the commercialization of products in cereal biotechnology.

# Chapter 7

# Cereal Biotechnology

The value of cereal crops can be improved in two main ways. Finally, the quantity of crop produced can be increased and secondly, the quality of the harvested grain can be enhanced. Linked to the question of quality is also the improvement of safety. These two approaches are not entirely independent. Improvement in the productivity of cereal varieties can be associated with improved quality of the grain. Removing constraints imposed by biotic stress (the impact of pests and diseases) is an attractive option for improving the productivity of cereal crops.

The introduction of genes conferring resistance to major pests and diseases provides an opportunity to improve productivity by removing the losses associated with the specific disease targeted. The introduction of herbicide resistance into cereal crops allows a reduction of losses associated with competition from weeds. Both these approaches may lead to associated improvements in grain quality. Freedom from weeds can reduce or eliminate the problem of weed seed contamination in cereal grain. Removal of disease constraints may result in improved grain quality avoiding the losses in quality associated with the presence of disease organisms in the crop.

Disease often result in reduced grain size and associated quality deterioration. The quality of cereal grain may be improved either in a nutritional sense or by improving the processing properties of the grain. Most conventional plant breeding has addressed the need for appropriate processing qualities in new cereal varieties. Biotechnology may allow more emphasis to be placed on novel alterations of nutritional quality. Molecular markers can be used to improve the efficiency of cereal breeding programs aiming to improve the value of cereal crops.

Molecular analysis may also be applied in fingerprinting or identification of cereal genotype with more immediate potential for improvement of cereals. For example, molecular analysis of genotypes can be used to monitor seed purity and identity prior to planting and to characterize grain lots in trading and processing. The composition of cereal-based foods or products can be monitored to ensure authenticity of labelling. These applications of biotechnology can have an almost immediate impact on the quality and value of cereal production. Genetic engineering offers the possibility of going beyond these short-term outcomes of biotechnology applications to the generation of more novel cereals with increased value in the longer term. In the sections that follow, the main benefits of genetic engineering are summarized under the following headings:

(i) Productivity

(ii) Product quality

(iii) Safety

## Weed Control

The control of weeds in cereal crops may have a major influence on grain yields but usually has a much lesser impact on grain quality. This is because the major effect is in early crop growth impacting more on grain number than size. Late weeds are an exception. Generally, contamination of seed crops with weed seeds is likely to be the major quality defect.

### Classes of Herbicides and Available Resistance Genes

Resistance genes are available for many different classes of herbicide. The major groups of herbicides and the genes available for conferring resistance to these herbicides are listed in Table 7.1.

The most attractive herbicide resistance genes for introduction into cereals are those that confer resistance to herbicides that are considered safe in the environment. Herbicides with low mammalian toxicity and little or no other environmental problem may be attractive alternatives to the more specific herbicides currently in use. The development of transgenic cereals with resistance to appropriate herbicides may facilitate the reduction in use of less desirable herbicides in agriculture and food production.

**Table 7.1. Herbicide resistance genes**

| *Herbicide* | *Mode of action* |
|---|---|
| Glyphosate | Inhibits 5-enolpyruvyl shikimate-3-phosphate (EPSP) |
| Sulphonylureas | Inhibits acetolactate synthesis (ALS) |
| Imidazolinones | Inhibits acetolactate synthesis |
| Triazalopyrimidines | Inhibits acetolactate synthesis |
| 2-dichlorophenoxyacetic acid | Auxin action |
| Phosphinothricin | Inhibits glutamine synthesis |
| Atrazine. | Inhibits electron transport in photosystem II |

### Problems of Escape of Herbicide Genes to Weeds

A major risk associated with the production of transgenic cereals with herbicide resistance is the possibility that new weeds may result either from escape of the gene into other plants or by the transgenic cereals themselves becoming weeks. The production of transgenic plants with resistance to herbicides that have unique modes of action is highly desirable. Multiple herbicide resistance may arise if genes target biochemical pathways that are associated with the action of several classes of herbicide. The problem of escape of herbicide resistance genes from cereals is likely to be a more serious issue for species that are out-crossing rather than for those that are predominantly or exclusively self-pollinating.

In some species such as rice, weedy various have developed as a result of current agricultural practices. The introduction of herbicide-resistant rices could lead to the development of a further class of weedy rices base upon their herbicide resistance if appropriate agricultural particles are not adopted in association with the new varieties. This could require production to be limited to specific regions and to include the rotation of herbicides or varieties. Despite these limitations, herbicide-resistant cereals should provide enormous advantages in the enhancement of cereal productivity.

## Disease Resistance

Disease may seriously reduce grain quality. Fungal disease may have a large impact on grain quality (especially grain size) because they may active on the leaves during grain filling or directly infect the head. Insect pests may reduce yield and grain quality. Post-harvest damage from insects can be a major problems.

Mycotoxins resulting from fungal growth on the grain are a serious safety issue for grain from some environments. Development of pest- and disease-resistant cereals provides a major opportunity for enhancing cereal productivity. In many environments single diseases may associated with very serious losses in grain yield.

Breeding resistant varieties has been a major strategy used in increasing cereal yields. Transgenic cereals with high levels of disease resistance may extend the options available from conventional plant improvement. Resistance to a wide range of biotic factors may be engineered using appropriate genes. Resistance to viruses, bacteria, fungi, nematodes and insects has been reported.

### Manipulation of Expression of Native Genes for Disease Resistance

Classical cereal breeding has involved the combination of disease-resistant genes from different sources to produce commercial varieties with effective disease resistance. Molecular markers or direct analysis for the presence of the required gene are now used to improve the efficiency of selection of disease-resistant lines in breeding. One option for the engineering of cereals with disease resistance is to manipulate or enhance the levels of expression of genes already present in the genome. Specific options include the expression of defence genes using constitutive promoters such that the defence gene product was always produced by the plant regardless of the presence of a specific pathogen. Promoters induced by the disease are also an important option. This approach allows defence gene products to be produced by the plant only in response to attack by the pest.

### Novel Genes for Disease Resistance

Novel genes from other plants or non-plant sources may provide durable resistant genes for use in cereals. Examples have been described for bacteria, fungi, nematodes and insects. The use of virus-derived sequences for breeding virus-resistant plants has been a notable success in many species.

### Environmental Impact of Disease Resistance

Transgenic plants with resistance to pests and disease may have a significant impact on the environment. The risks should be similar to those of traditional resistance breeding. However, if the transgenic strategies prove dramatically more effective they could seriously deplete population of plant pathogens and even lead to the extinction of highly specific pest organisms. Evaluation of these risks is important is successful applications of transgenic technology.

**Table 7.2. Some novel pest- and disease-resistant genes of potential value in cereals**

| *Type of protection* | *Gene* | *Source* |
|---|---|---|
| Virus | Coat protection | Barley yellow dwarf |
| | Coat protection | Maize chlorotic mottle |
| | | Maize chlorotic dwarf |
| | | Maize chlorotic mosaic |
| Bacterial | Chitinase | Various |
| and Fungi | Glucanase | Various |
| Insects | Bt toxin | *Bacillus thuringiensis* |

## Improved Nutritional Properties

### Importance of Cereals in Human and Animal Nutrition

Cereals are a very important part of human diets. The three major species, wheat, maize and rice, account for a large proportion of the calories and protein in human diets. The importance of cereals in the food chain is also protein in human diets. The importance of cereals in the diets of animals. The major constitutes of cereals are the carbohydrates and proteins. Other grain components such as lipids and vitamins may be of great significance in human nutrition because of the large contribution of cereals to the diet.

Biotechnology provides new options for manipulation of the nutritional properties of cereal grains. The carbohydrates of cereals include the simple sugars, the more complex oligosaccharides such as fructans, storage polysaccharides of the grain (starch) and the cell wall polysaccharides, all of which are of nutritional value. All of these carbohydrate components are potential targets for manipulation in improvement of cereal quality. (For example, sugar beet has been transformed to produce fructans). Benefits that may result include reduced cariogenic bacteria (dental health), lower energy value and stimulation of beneficial bacteria in the colon.

The sugar content may also influence the quality of the grain for various products. Fructans may be considered to be important to human nutrition because of their possible role as soluble fibre. Starch, as the major component by weight of the grain, may have a great impact on nutritional quality. Resistant starches (not digested in the gut) may be considered critical in influencing the incidence of certain human diseases, such as heart disease. The cell wall polysaccharides may also be important as either soluble or insoluble fibre, depending on the composition of the polysaccharides in the cereal product.

Soluble fibres may reduce the risk of hear disease while insoluble fibres contribute to reduced risk of colonic cancers. Cereal proteins are not well balanced in amino acids required in a nutritionally balanced diet, and genetic engineering may provide opportunities to improve the balance of essential amino acids in cereal-based diets. The lipids in cereals are generally of limited importance in human nutrition but may be important in animal diets. The manipulation of iron levels in cereals through the introduction of haemoglobin illustrates the potential application of biotechnology to enhancing the nutritional value of cereals.

### Animal Foods

The major requirements of animals differ depending on whether they are monogastric or ruminant animals and the potential of biotechnology to improve the nutritional values for these two classes of animals differs. Ruminants have a much greater capacity to digest cereal fibre effectively. Soluble fibre components may also be important. For example, the β-glucans of barley limits the use of this cereal in the diet of chickens because of its adverse impact on nutrient utilization.

### Aquaculture Foods

The increasing shortage of seafoods (declining fish stocks in the oceans and increasing human population) indicates great potential for enhanced use of cereals in aquaculture diets used in fish farming. Cereals provide a very cheap option and if biotechnology can be used to enhance the nutritional value of cereals as a component of aquaculture diets we can expect wide-scale use of cereals for the production of aquaculture products. Cereals may have an important role as a binder in aquaculture feeds. Improvement of protein and lipid composition by genetic engineering may produce more useful cereal aquaculture feeds.

## Improved Processing Properties

The quality requirement of cereal processors may be complex as indicated for barley in Table 7.3. This table lists a few of the characteristics defined as requirements of a barley for use in malting and brewing. Establishing opportunities for quality improvement requires a knowledge of the processes used to convert cereals into end products.

**Table 7.3. Barley quality characteristics required for brewing**

| *Character* | *Requirement* |
|---|---|
| Grain colour | Bright, white aleurone |
| Grain size | Plump grains, 90% above 2.5 mm, as little as possible below 2.2 mm |
| Protein content | Optimum 10.5%-11.5% (dry basis) |
| β-glucan content | Low- maximum and minimum not defined |
| Husk | Minimum required for brewing |
| Dormancy | As little as possible (but no pre-harvest sprouting) |
| Rate of modification | As fast as possible (steeping and germination maximum 110h) |
| Malt extract | As high as possible (only too high if protein on husk becomes limiting) |
| Malt enzymes | As high as possible (minimum requirements for diastatic power and β-glucanase) |

## Introduction to the Range of Processes that might be Manipulated Genetically

The processing of cereals involves a wide range of techniques with very different raw materials requirements. Several of the major processes of cereal processing will be described here in an attempt to identify the major opportunities for biotechnology to be applied to improving cereal processing quality.

### *Milling*

The milling of cereals involves processes that are dependent substantially on the anatomical structure of the grain. The potential for single genes to be manipulated in ways that enhance milling quality my be limited because of the large number of grain characteristics contributing to milling performance. The shape of the grain and adherance of the various outer layers are of great importance. Hardness is also a key attribute in milling. Grain colour is a key quality attribute that is influenced by the milling processes. The levels of pigments and the size of particles generated in milling both influence colour. The contamination of endosperm fractions

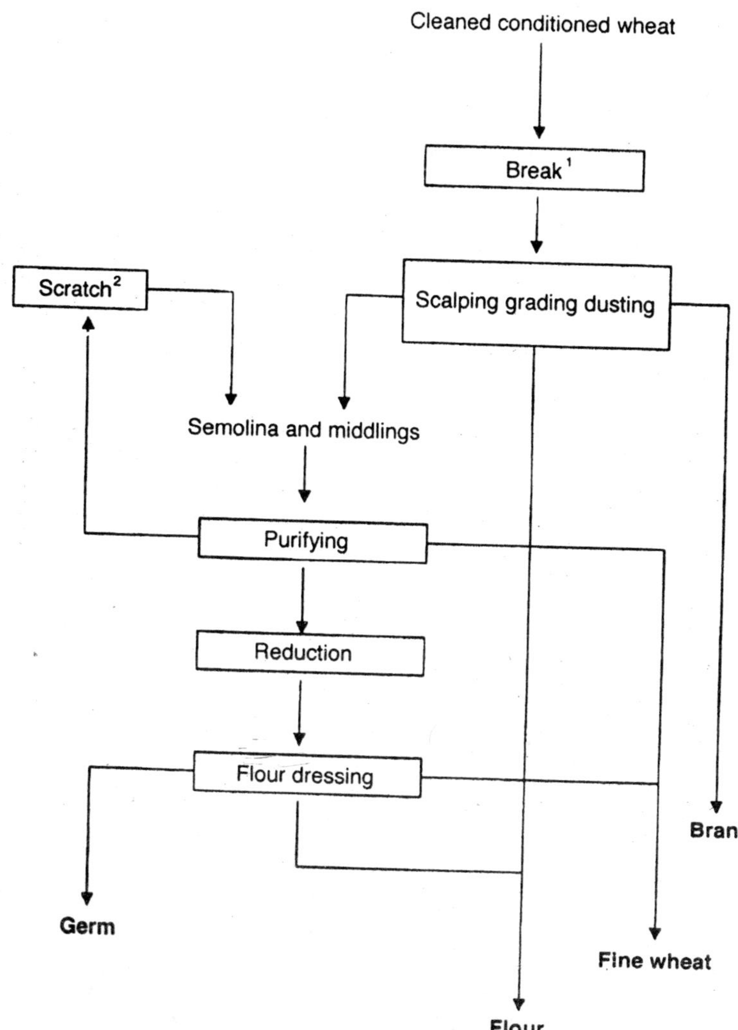


*Fig. 7.1. Schematic of flour milling.*

with more highly coloured outer layers such as bran are key determinants of colour.

## *Baking*

Many components of cereals such as wheat are important in baking quality and are obvious candidates for the application of biotechnology. The requirement of modern high-speed plant bakeries are for a very consistent quality. Characteristics such as dough stickiness associated with some alien (non-wheat) sources of disease resistance in wheat are serious problems in these high-capacity facilities. The major components of wheat all contribute to baking quality and are thus targets for genetic improvement of baking quality. Proteins are essential for the viso-elastic properties of wheat doughs. Starch and cell wall polysaccharides (e.g. pentosans) also influence baking quality. The breakdown of starch by amylases is a key process in baking. The pentosans of the cell wall also have a significant influence on loaf quality.

### *Malting*

Malting is the first step of processing grain for use in brewing and distillation. The malting of cereals, most specifically barley, is a process of germination. The rate of germination and the changes in the composition of the barley during malting are potentially important targets for biotechnology. The breeding of malting barley varieties focuses on several major quality characteristics associated with the malting process. Enzymes are involved in the breakdown of cell walls during malting. The β-glucanases expressed during germination are essential to ensure that the levels of β-glucan in the malt are low. High malt β-glucan levels contribute to high wort viscosity, poor wort and beer filtration and potential hazes in finished beer.

### *Brewing*

The brewing of beer from malt requires specific malt specifications that are potentially able to be manipulated using biotechnology. Diversification of beer styles and markets are imposing divergent raw material requirement for the different beer styles.

Sufficient levels of starch degrading enzymes in the malt are necessary to ensure breakdown of starch to fermentable sugars during brewing. The amount of starch (adjunct) added in the form of rice or maize is a key determinant of the level of starch degrading enzyme required in the malt. The relative levels of different starch degrading enzymes are important in determining the nature of the substrate for fermentation and the sugar, alcohol and oligosaccharide content of the beer. The alcohol and residual sugar content (sweetness) are influenced by the levels of fermentable sugars and the non-fermentable oligosaccharides contribute to the taste (mouthfeel).

### *Distilling*

Distillation is a process where the product quality may be less dependent on the raw materials than many other processes and may be a less important target for biotechnology application in relation to the cereal raw material. However, the quantity of starch available for fermentation might be enhanced.

### *Extrusion*

Extrusion (production using high temperatures and pressures) of cereal products is an increasingly important process in the production of. a wide range of products including snack foods, breakfast cereals and pet foods. The processing properties required are complex but may be enhanced by the application of biotechnology.

### **Wheat Utilization**

Wheat is used for a wide range of products with differing quality requirements. The importance of different wheat grain components and characteristics depends on the ultimate end use product. The protein quality is much more important for products such as breads than for cake sand biscuits. Enhanced levels of desirable high molecular weight glutenins may be desirable in wheat for use in breadmaking. Manipulation of starch synthesis and starch properties may be more important in products such as noodles. Colour may be controlled by a small number of genes and has differing importance. For example, the yellow pigments in durum wheat are considered highly desirable while some noodle products require very white flour.

The improvement of specific attributes using biotechnology needs to target characters specific for particular end uses. Improvement of the value of wheat for this wide diversity of uses requires the matching of wheat characteristics to specific end product requirements. For example, wheat with different combinations of protein and hardness are better suited to particular end uses. However, some combinations of characteristics will not be optimal for any major end use. Genetic improvement of wheat quality needs to address targets relevant to the end use characteristics of wheat from a particular environment or region. For example, selection for specific starch metabolism mutants or engineering of improved starch qualities (such as starch-pasting properties) may be important in regions producing noodle wheats

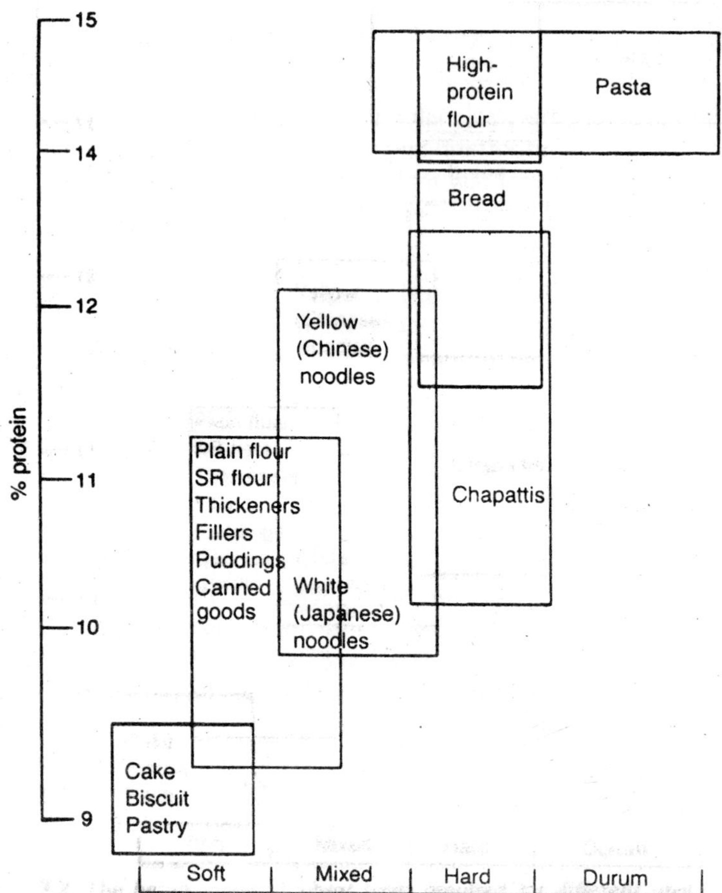


*Fig. 7.2. The broad range of wheat types required for different uses.*

while storage protein modification may be more appropriate in regions producing bread wheats.

## Beer Production

The production of a specific product such as beer can be the basis for analysis of opportunities for application of biotechnology. Some key attributes for malting and brewing have been described above. The efficiency of this process may be influenced by the composition of raw materials (especially the cereals) used and the type of beer to be produced. Changing composition of barley by genetic engineering might alter performance in any of the many steps in the process of beer production. The production of proanthocyanidin-free barley to reduce haze and improve the shelf

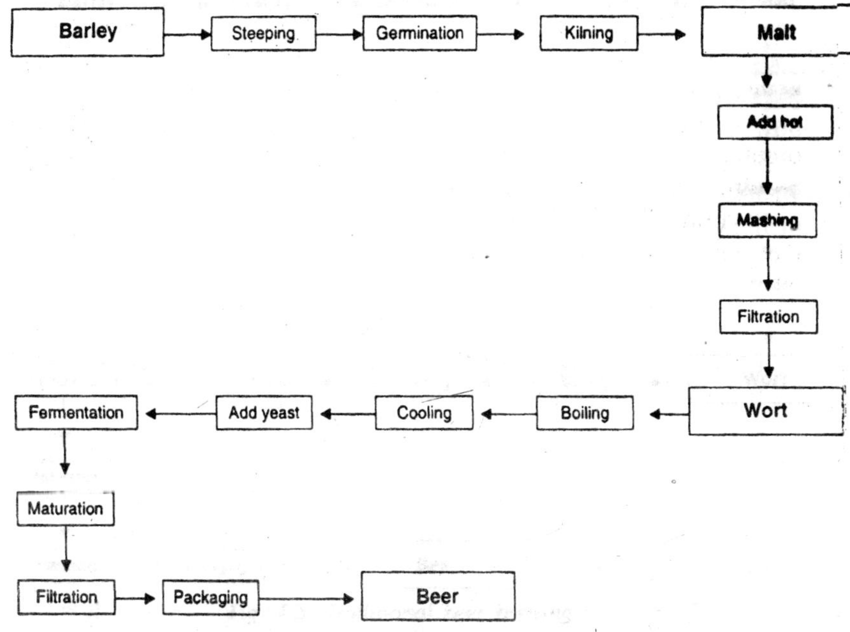


*Fig. 7.3. Traditional beer brewing.*

life of beer in cold storage has been achieved using mutants and may be more precisely controlled using genetic transformation to block specific steps in the pathways leading to proanthocyanidin formation. One complication of this change may be that the absence of proanthocyanidins in the wort may result in reduced protein precipitation during boiling.

The overall result can be a decline in product stability due to increased levels of protein in the beer. This illustrates the need to understand all the interactions during processing before implementing changes to novel cereal raw grains. Molecular markers have been developed for many malting quality attributes and these may accelerate the development of new malting quality barley varieties for use in beer production.

Genetic engineering of barley to improve cell wall breakdown during malting and brewing and lipids contributing to beer flavour have been the subject of active research. Modifications of starch properties and metabolism and protein composition may also be important. A fundamental limitation is that extract levels are limited by the need to preserve a minimum amount of husk to act as a filter bed in many traditional brewing processes. Grains with extremely high proportions of endosperm necessary for extreme extract levels will have insufficient husk. New filtration technologies could over come this limitation allowing the development of very high extract barleys. Reliance on enzymes in the malt for starch breakdown during mashing and the tissue of yeast nutrition also requires that a minimal protein level be maintained. Genetic engineering could allow a higher proportion of proteins in the malt either to contribute useful enzyme activity or support yeast nitrogen nutrition.

## Improved Cereal Quality Control

The value of cereals can be improved by better specification of the identity and composition of cereals in marketing. The major contributions to achieving this type of value addition are quality assurance programs and enhanced tools for analysis of products for chemical and microbial contamination and for genetic identity and purity.

### Chemical and Microbial Purity

The strict application of quality assurance principles can ensure cereal safety and value. New technologies allow the use of ELISA tests to establish the level of pesticide residues and DNA-based analysis of microbial contaminants.

### Genetic Purity

The identity of a cereal genotype defines many of the quality characteristics such as protein content and grain size. This can define the processing value and optimal end use of a parcel of cereal grain. Protein content can currently be measured rapidly in the field using near intra-red reflectance (NIR). The development of rapid genotyping methods would find wide application in the cereal industry and allow a relatively complete objective description of samples for commercial valuation. Distinction of high-value noodle wheats from visually similar wheats of low noodle quality, and distinction of mating and food barleys with similar appearance, are good examples of the areas of potential application of this technology. Genetic purity and level of admixture are also important attributes to assess in commercial trading because of the financial incentive to add low-value grain to parcels of very high value but visually similar genotypes. Rapid DNA extraction from grain is a key technical requirement for successful application of these methods.

## Future Prospects and Limitations

Biotechnology is likely to have a major impact on the value of cereal production both by increasing productivity and by improvements in product quality. The improved productivity is likely to result initially from the removal of biotic stress constraints associated with major pests and diseases. Herbicide resistance is an option that is likely to be able to achieve early adoption and success. Improvements in grain quality are likely to be generally more difficult to achieve.

The major attraction of biotechnology is the possibility of introducing totally new or novel characteristics into cereals that will result in products with characteristics outside the range of those currently available. A major limitation to the introduction of such characteristics is the requirement of cereals to be compatible with existing processes of cereal food production. Market resistance to products requiring new processing techniques will come from the large investment that may be required to develop new processing facilities. Improved method of quality control and analysis of product identity and purity will enhance the value of cereal products. Adverse consumer attitudes are also a significant risk if transgenic products are not well designed and marketed.

# Chapter 8

# Transgenic Potato

Potato is the world's fourth most important food crop behind wheat, rice and maize. Over the last three decades potato production has grown faster than any other food crop except wheat (FAO). Glennon (2000) states that, agriculturally, in the eyes of the developing countries, no other crop has more production potential, since yield potential is still largely under exploited. In developing countries potato is also seen as a candidate for resolving domestic production problems. More than one billion individuals (50% of these in developing countries) now eat potato and as little as 100 grams supplies 10% of the recommended daily calorie allowance children. The same amount provides about 10% of essential vitamin intake (e.g. thiamine, niacin, folate) and 50% in the case of vitamin C.

In 1998 global production was around 290 million tones with 30% of production in developing countries. In the European Union of 15 member states production totals 50 million tones (MT) and is dominated by Germany (11.3 MT), the United Kingdom (6.6 MT), France (6.5 MT) and the Netherlands (6.0 MT), whilst in consumption terms Ireland still has the highest intake at 140 kg per capita. What are the most important subjects demanding global attention in potatoes? Collins (2000) indicates that in developed countries the most important include:

1. Disease control strategies for the late blight, bacterial wilt, ring not, nematodes and threats to the available of appropriate chemicals to control these pests and diseases;
2. Processing and marketing;
3. Seed tuber quality and health;

4. Genetically Modified Organisms (GMOs) and related issues (acceptance and use of biotechnology, ownership of intellectual property, freedom-to-operate); and
5. Genetic resources for future use.

**Table 8.1. Top ten world potato production 1998**

| | *Production (million tones)* | | *Production (million tones)* |
|---|---|---|---|
| China | 45 | India | 19 |
| EU 15 | 48 | Ukraine | 17.5 |
| Russia | 37 | Belarus | 10 |
| Poland | 26 | Turkey | 5.3 |
| USA | 21 | Canada | 4 |

Collins also point that resources in public institutions in developed countries, which traditionally addressed these problems, are shrinking at an alarming rate. Solutions might include bilateral country collaborations on common priorities, new partnership with the private sector, regional/global cooperation on more widespread issue. More than 40% of the world's potatoes are grown in developing countries and this is expected to increase. Within both developing and developed countries there is also a trend towards a decline in fresh consumption and a continued rise in process utilization.

**Global Initiative on Late Blight (GiLB): a Model for Collaborative Research**

An example of a global initiative to resolve a key biological problem facing potato is the Global Initiative on Late Blight (GiLB). This was formed following the relation that the magnitude of the late blight problem requires a global strategy. Late blight is caused by the fungus *Phytophthora infestans*. The pathogen is showing increasing resistance to chemical used in developing countries (and new chemicals are subject to increasing safety regulations). It also travels easily, providing access to environments where variability can be increased through sexual hybridizations.

Organizations such as the International Potato Center (CIP) are dedicated to improve the health and well being of disadvantages populations in the developing world and appreciate that there is a role for genetic modification (GM) technology in such initiatives. CIP is playing a key role in delivering links between private and public center research and has set up a global strategy for the uptake and development of genetically engineered potatoes when and where appropriate. The initiative is called potato GENE (Genetic Engineering Network) and currently involves scientists from 11 developing and three industrialized countries. Four priority areas have been focused upon, which actually reflect key issue for the developed as well as developing countries with regard to GMOs:

1. Strategies for addressing public concerns;
2. Seed systems and intellectual property;
3. Gene flow and effects on non-target organisms; and
4. Health and food safety issues.

The objective is to provide the scientific information necessary for informed choice. The GM debate in general is opening up and considerable transparency is emerging though initiatives driven by scientists themselves.

## Potato Breeding and a Role for GM Technology

Potato breeding started around the beginning of the 19th century but its complex tetraploid genetics means that targeted breeding is a time-consuming exercise. Comparatively speaking potato has a narrow genetic base, which, at least in part, contribute to slower progress in crop improvement than with other major crop species. Attempts to introgress genes from wild relatives has met with some success for disease resistance traits but undesirable side effects from hybridization events mean that periods of ca. 12 to 15 years have been required for cultivar development.

Since back-crossing to remove undesirable effects is not an easy option for potato, the approach of improving existing cultivars using gene transfer or genetic engineering technology has been an attractive proposition. Realistically, this approach is not meant to replace the role of the plant breeder, but complement it by providing additional tools to modify potato genomes for environmental and commercial benefit. From an environmental perspective GM potato poses few risks as pollen movement is extremely limited (separation distances for experimental release of nine metres are acceptable in the UK) and there are no wild relatives with which cultivated potato can outcross.

### Potato Transformation

The traditional method of potato genetic transformation employs *Agrobacterium*-mediated systems. Whilst transgenic potato is often used as model systems to assess the roles of specific gene(s), cultivars such as Desiree tend to dominate the 'transgenic' literature as transformation efficiencies are high. However, Dale and Hampson (1995) examined transformation efficiencies in 34 potato varieties using a tuber disc protocol, showing that only half of the cultivars regenerated. From those that could be regenerated all but one produced transgenic plants.

Some cultivars which did not regenerate from tuber discs did so from leaf and internode segment. It follows, then, that to deliver a commercial product which is true-to-type but which has the desired level of transgene expression and trait modification may not yet be a facile operation with some genotypes. Belknap et al. (1994) transformed Lemhi Russet and Russet Burbank with constructs containing GUS, ClaSP (tyrosine-rich arylphorin) or a gene encoding abacterial lytic peptide and showed that the highest potential for deviation from typical performance occurred in yield and tuber size gradings. The frequency of off-types varied between 15 and 80%, depending on cultivar, but off-types were not always apparent until plants were grown in the field. Of the lines transformed with GUS, less than 50% produced seed tubers under field conditions. Field trialling is clearly imperative to gain a real insight into compositional, biochemical, phenotypic and agronomic performance.

## Somaclonal Variation

Potato is potentially a good model crop for selection of improved lines generated through somaclonal variation from novel variants can arise. Sexual crosses are not always possible in potato due to sterility problems or lack of flowers and, as already stated, the genetics of tetraploid inheritance is problematic. Potato is easily regenerated in tissue culture (although as stated previously some cultivars are more recalcitrant than others) and is vegetatively propagated from tubers. However, somaclonal variation may produce undesirable effects following targeted genetic transformation events, thus modifying phenotype and agronomic performance independently of any effects induced by insertion of the target gene(s). The *in vitro* regeneration process required to produce GM potato lines involves: (a) establishing de-differentiated cells from tissue or organ culture under defined conditions; (b) proliferation for a number of cell generations; and (c) subsequent plant regeneration under *in vitro* conditions.

Somaclonal variation in regeneration plants is generated during *in vitro* culture stage and particularly during de-differentiation. This is accompanied by increased frequency of chromosomal abnormalities with time in culture. Genetic changes also occur in plant tissues and cells *in vivo* due to mutations, endoreduplication, chimeras etc. Genetic variation in plants regenerated in vitro can therefore be derived from *in vivo* and *in vitro* events. The contributions of *in vivo* and *in vitro* modifications are dependent on parameters including genotypic background, culture conditions, etc. Somaclonal variation is uncontrollable and unpredictable in nature and most variation is of no apparent use. However, stability of any useful somaclones produced may not be a problem.

Morphological changes observed range from gross abnormalities to minor and more subtle modifications. There is distinct genotypic variation in the frequency of somaclonal variants that might arise. Thus selecting GM Potato lines for commercialization which have the desired impact but in which other traits are not significantly modified by the tissue culture process will require the production of several hundred independently transformed lines and full and effective field selection using criteria that breeders would normally impose. This will be in addition to the testing of a range of constructs, promoters, targeting sequences, etc. where relevant. Compliance, in risk assessment exercise, with the need to demonstrate 'substantial equivalence' of a GM line with the parent from which it is derived should take into account compositional variation that might be induced by somaclonal variations in species such as potato, and not only from the expression or insertion of target genes.

The concept of substantial equivalence is that the GM line to be marketed should be compositionally the same as the parent line from which it was derived, but with the exception of any modification expected by inserting the gene of interest. This has to be assessed using several growing sites over more than one year. From a personal perspective the value of substantial equivalence should be combined with knowledge of natural genetic variation in the compositional status of cultivars already in production and on sale. For this reason extensive databases which clearly demonstrate the ranges of metabolite concentrations that might be expected in species of crop plants will have great utility. A scenario can be envisaged in which the composition of a GM line may differ from its parent but fall well within the range expected of cultivated potato.

## COMMERCIAL APPLICATIONS OF GM POTATO CROPS

Global sales for transgenic crops were estimated at $75 million in 1995, increasing to $235 million in 1996 and $670 million in 1997. In 1999 the estimated value was approximately two billion dollars (a 30-fold increase in five years). In descending order of cropping area the ranking order for GM crops is soybean (54% of total), maize (24%), cotton and canola (rapeseed; 9%), potato, squash and papaya (1% or less). The USA has the highest proportion of GM crops with 74% of the world total, followed by Argentina (15%), and Canada (10%). Herbicide-tolerant crops occupy 71% of the area grown, insect-resistant (*Bt*) crops 28%. Romania and the Ukraine have grown introductory areas of *Bt* potato (less than 1000 ha) but the majority of commercial potato crops have been grown in North America. No potato crop currently has European Union approval for commercial release into the environment. This chapter is primarily aimed at the commercial applications of GM potato. When the chapter was conceived the only company to have this crop in the commercial market place was Monsanto, a St Louis-based agricultural biotechnology and herbicide company of which the Pharmacia Corporation now owns 85%.

Monsanto markets GM potato through its NatureMark® biotechnology unit. The potato lines developed by NatureMark® will be outlined later. According to a Reuters report on 21 March 2001, the Monsanto Company stated that it would no longer be marketing genetically modified potatoes as part of a streamlining of its biotechnology crop portfolio to focus on wheat, corn, soybeans and cotton. This decision was not doubt influenced by statements from some leading processing and fast-food outlet companies that consumer concerns over GM potato products had affected their GM purchasing policy.

Since the processing market is the major one for potato in North America, the GM potato business becomes non-viable as a result. Today's reliable, cheap, potatoes are a testimony to the ingenuity of scientists, agronomists and farmers over the past 100 years, and a tremendous success story by any standards. However, even with all the advances in agricultural science some 40% of potato harvests are lost worldwide to diseases, pests and weeds, a figure which might reach 75% if it were not for crop protection products. NatureMark® commercialized its potato GM lines through the NewLeaf™ brand and the first-generation targets were aimed at resolving some of these key production issues. The lines developed are described below.

**Insect-resistant potato: NewLeaf™**

NatureMark® potato lines named NewLeaf™ confer resistance to Colorado potato beetle (CPB; *Leptinotarsa decemlineata* Say). Colorado beetle is a primary pest of North America and elsewhere, larvae emerging from egg masses about one week after deposition then feeding on foliage. Mature larvae leave the plant and pupate in soil to emerge as adults one week later and begin feeding on foliage yet again. NewLeaf™ potatoes were commercialized in the USA and Canada after more than ten years of development work in laboratories, glasshouses and fields across North America. The potatoes were transformed with a gene, which encodes for the Cry3A protein from the bacterium Bacillus thuringiensis (var. tenebrionis) using the constitutive 35S CaMV promoter. The Cry 3A protein is part of a family of proteins used for more than 30 years by organic producers, home gardeners, etc.

The product has been endorsed by the World Health Organization (WHO) and other regulatory agencies throughout the world. The protein affects directly only the target pest, Colorado potato beetle (CPB) and has no effect on other insects, mammals or wildlife. Growers have reduced insecticide usage on average by 42%. NewLeaf™ was approved in the USA in May 1995. The relevant agencies in the USA, the Food and Drug Administration, (FDA), United States Department of Agriculture (USDA) and Environmental Protection Agency (EPA) have approved the product based on substantial equivalence to other Russet Burbank potatoes. The product has also been approved by Health Canada, Agri-Food Canada and Agriculture Canada. Japan and Mexico have also approved the use of NewLeaf™ potato. The first generation of NewLeaf™ potatoes included cultivars Atlantic, Superior and Russet Burbank and all of these NewLeaf™ varieties have been completely de-regulated by the FDA, USDA and EPA in the USA.

A range of transgenic lines were selected with superior agronomic characteristics in additions to CPB resistance. Cultural management guidelines were also developed for growers. Out of the programme arose NewLeaf™ 6 Russet Burbank which is high yielding (earlier tuber bulking) and which produces a high percentage of quality grade potatoes classified as US and Canadian number ones. This has required optimization of row spacings and development of appropriate fertilizer regimes. NewLeaf™ 6 also has extended dormancy and shows improved long storage characteristics. Since the potato is completely protected from CPB it appears to need no additional protection from this pest.

Plants modified to express insecticidal proteins from *Bacillus thuringiensis* (referred to as Bt-protected plants) are believed to provide a safe and highly effective method of insect control. *Bt*-protected corn, cotton, and potato were introduced into the United States in 1995/1996 and grown on a total of approximately ten million acres in 1997, 20 million acres in 1998, and 29 million acres globally in 1999. These crops provide highly effective control of major insect pests such as the European corn borer, southwestern corn borer, tobacco budworm, cotton bollworm and pink bollworm, in addition to CPB. They reduce reliance on conventional chemical pesticides and appear to provide notably higher yields in cotton and corn.

**Table 8.2. NewLeaf™ commercial adoption rate**

| | |
|---|---|
| 1995 | 1,800 acres of commercial production<br>54 M lbs of raw product used for tablestock and French fries. |
| 1996 | 10,000 acres of commercial production.<br>300 M lbs of raw product used for tablestock and French fries. |
| 1997 | 30,000 acres of commercial production.<br>900 M lbs of raw product used for tablestock and French fries. |
| 1998 | 48,000 acres of commercial production.<br>1.4 B lbs of raw product used for tablestock and French fries. |
| 1999 | 55,000 acres of commercial production.<br>1.65 B lbs of raw product used for tablestock and French fries. |

The estimated total net savings to the grower using *Bt*-protected cotton in the United States was approximately $92 million in 1998. Other benefits of these crops include reduced levels of the fungal toxin fumonisin in corn and the opportunity for supplemental pest control by beneficial insects due to the reduced use of broad-spectrum insecticides. Insect resistance management plants are being implemented to prolong the effectiveness of these products. Extensive testing of *Bt*-protected crops has been conducted which establishes the safety of these products to humans, animals and the environment.

Acute, sub-chronic, and chronic toxicology studies conducted over the past 40 years have established the safety of the microbial *Bt* products, including their expressed insecticidal (Cry) proteins, which are fully approved for marketing. Mammalian toxicology and digestive fate studies, which have been conducted with the proteins produced in the currently approved Bt-protected plant products, have confirmed that these Cry proteins are non-toxic to humans and pose no significant concern for allergenicity. Food and feed derived from *Bt*-protected crops, which have been fully approved by regulatory agencies, have been shown to be substantially equivalent to the food and feed derived from conventional crops.

Non-target organisms exposed to high levels of Cry protein are virtually unaffected, except for certain insects that are closely related to the target pests. Because the Cry protein is contained within the plant (in microgram quantities), the potential for exposure to farm workers and non-target organisms is extremely low. The cry proteins produced in *Bt*-protected crops have been shown to degrade rapidly when crop residue is incorporated into the soil. Thus the environmental impact appears to be negligible. The human and environmental safety of *Bt*-protected crops is further supported by the long history of safe use for *Bt* microbial pesticides around the world. For several decades the primary control strategy for CPB has been the use of chemical insecticides and about 22 active ingredients are registered, in Canada for example, for this purpose. However, restrictions in the modes of actions of insecticides coupled with repeated applications have led to resistance development in most commercial production regions in Canada.

The entry of new products with novel modes of action such as *Bt* toxin and products such as cyromazine and imidacloprid give more grower flexibility and reduce the potential for resistance development. Other insecticides are under development. Insecticide with *Bt* protein as the active ingredient is primarily effective against larvae only. Crop rotation is an important control strategy for growers too, as might be biological control using insect-destroying fungi such as *Beauveria bassiana* and beneficial nematodes such as *Steinemema carpocapsae*. However, the most spectacular development has been with *Bt* transgenics. Tests on Prince Edward Island and elsewhere in North America have shown that the technology is effective against both adults and larvae.

Potato plant mixtures of 70% GM and 30% non-GM appear to be as effective in controlling CPB beetle numbers as GM monocultures. Indeed, border rows of transgenics can reduce the number of colonizing beetles entering a field. CPB appeared in Europe in the 1920s and since 1980 has occurred in practically the entire European continent with the exception of the UK and Scandinavia. Quarantine actions are implemented only in the UK. Recent work has evaluated potato tuber moth resistance in tubers of transgenic potato lines expressing an alternative Bt protein encoded by the *Bt-cry5* gene. The potato tuber moth is the most destructive pest of potato in tropical and sub-tropical countries. Larvae attack both foliage and tubes in the field and in storage. Moth mortality was 100% in transgenic lines of cultivar Spunta using the constitutive cauliflower mosaic virus promoter.

**Virus-resistant potato: NewLeaf™ plus and NewLeaf™ Y**

Virus resistance in potato has been developed using a range of approaches and genetic constructs which include sequences of virus coat proteins, movement proteins, replicases, untranslatable sense or antisense RNAs, proteases, defective interfering RNAs, and satellites. Expression of ribozymes, a double stranded RNA-specific ribonuclease, antiviral proteins, a plant pathogen resistance gene and 'plantibodies' have also provided virus resistance for a review of approaches used to deliver virus resistance. NewLeaf™ Plus produced by NatureMark® is a high yielding Russet Burbank with combined CPB (*cry*3A expression) and potato leafroll virus (PLRV) resistance (produced suing the constitutive Figwort Mosaic Virus Promoter (FMV) within a construct designed to prevent virus replication).

PLRV can cause yield losses of as much as 50% and nearly all commercial varieties are susceptible to infection with worldwide losses estimated at 10%. PLRV also causes net necrosis (phloem cells affected), which greatly reduces the value of tubers for fresh and processing use. Freedom from such internal necroses provides a more consistent product and better financial returns per hectare by reducing processing costs in French fry and chip (crisp) industries and by helping to deliver improved seed quality. The transgenics are capable of reducing insecticide usage by up to 100%. NewLeaf™ Plus was approved in the USA for consumption in August 1997. The FDA and EPA in the US determined that the potato was as safe to eat as any other Russet Burbank. Large-scale agronomic trials were grown in the USA in 1998. NewLeaf™ Y cultivars Russet Burbank and Shepody have been developed with combined CPB and potato virus Y (PVY) resistances. PVY is considered one of the most damaging potato viruses because it causes economically significant yield depression. Severe infestations can reduce yield by as much as 80%.

The PVY coat protein gene used to generate resistance is also expressed using the FMV promoter and is more effective at PVY control than any insecticide programme allowing more sustained crop protection through reduced insecticide usage. Protection against PVY has reduced seed de-certification risk for seed growers and helped to maximize yields for commercial growers. Other benefits include improved processing quality and storage and higher tuber set in cultivars such as Shepody (more uniform tuber size distributions have been claimed through improved line selections). As far as NewLeaf™ Y is concerned the FDA and Health Canada completed their review in May 1999 and agreed they were safe for human consumption. The EPA, USDA and Canadian Food Inspection Agency determined that NewLeaf™ Y poses no concern for unreasonable effects on the environment or livestock. The process started with *ca.* 3,000 original clones for each with the first selection occurring after transplanting to soil and removal of off types.

True to type plants were then subjected to a range of tests including phenotypic analysis, efficacy tests against CPB, PLRV and PVY, susceptibility to inset, fungal and bacterial pests, evaluation of tuber yield, tuber appearance, quality and additional extensive field performance evaluation. Field evaluation took place over several years, in different environments and under various agronomic conditions with emphasis on potential environmental impact. From this process six clones were selected for commercialization. Rogan et al. (2000) analyzed key nutritional, quality and anti-nutritional components of NewLeaf™ Plus and NewLeaf™ Y lines to assess their substantial equivalence to the

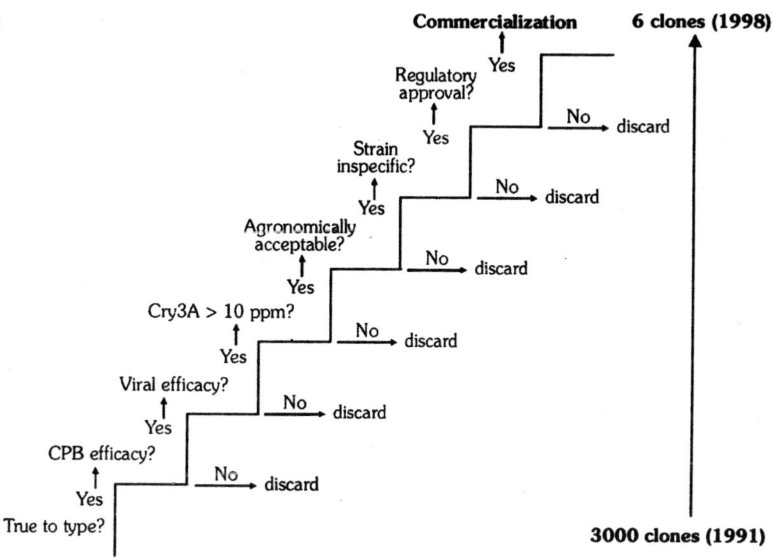


*Fig. 8.1. Steps to commercialization for NewLeaf™ Plus and NewLeaf™ Y CPB and virus-resistant potato clones.*

parent cultivar. Dry matter content, vitamin C, soluble sugar, soluble protein, glycoalkaloids, vitamin $B_6$, niacin, copper, magnesium, potassium, amino acids, fat, ash, calories, total protein and crude fibre were quantified. The data confirmed that tubers produced by insect and virus protected varieties were substantially equivalent to tubers produced by conventional varieties. Riebe and Zalewski (submitted for publication) surveyed insecticide usage on NewLeaf™ Plus potatoes in paired field comparisons with conventional Russet Burbank.

The work showed that insecticide use could be greatly reduced or eliminated with the transgenic material, providing significant environmental and economic benefit. Net savings to growers averaged \$212 and \$313 per hectare in 1998 and 1999, respectively. On the 140 ha monitored in 1998, NewLeaf™ Plus allowed growers to reduce insecticides and miticides by a total of 2,870 kg of active ingredient and 7,700 kg of formulated material as compared with an adjacent field of conventional crop. Data indicate that more than 500,000 kg of active ingredients could be eliminated annually in the Columbia Basin, USA, if NewLeaf™ Plus replaced all of the 35,000 ha of Russet Burbank grown in the region.

## Transferring Virus Resistance Technology to Developing Countries

In 1991 a transfer agreement was initiated between Monsanto, who donated virus resistance technology and transgenic know-how, and the Centre for Research and Advanced Studies (CRAS), a public Mexican organization. The project was brokered by ISAAA (International Service for the Acquisition of Agri-biotech Applications). Currently, varieties resistance to potato viruses X (PVX) and Y (PVY) are in seed increase and final stages of regulatory approvals. Other, involving triple resistance to PVX, PVY and PLRV are undergoing line selection prior to field testing. These potatoes may become the first approved transgenics that local scientists in a developing country have generated through gene transfer and product development. Within the Mexico's potato farming system the smaller, resource-poor farms purchase fewer certified, clean seed and most plant tubers harvested from a previous year. This result in a constant build up of virus in the stock which affects yields. PVX-PVY-PLRV-resistant varieties should decrease unit production costs on farms by *ca.* 30%. Virus resistance delivered through transgenics have obvious benefits but will need to be established within an improved framework of seed potato production and distribution within countries such as Mexico.

## Potential Risks Associated with Transgenic Virus Resistance

The development of transgenic virus resistance has raised concerns over potential interactions between viral transgenes or their products with viruses that infect the transgenic plant itself. Such interactions include genome recombination and transcapsidation (the encapsidation of one virus with coat protein of another virus). This has been shown to occur in some instances where several viruses infect the same plant. Risks are perceived since capsid protein influences viral transmission properties. Thomas et al. (1998) looked for evidence of interactions between PLRV-derived transgenes and virus to which the transgenic potato plants were exposed and infected. Over 25,000 plants and 400 lines were transformed with 16 different coat protein constructs and *ca.* 40,000 plants and 500 lines with seven different replicase gene constructs of PLRV.

Heterologous viruses found infecting the plants were screened for modifications in transmission characteristics, host range, symptoms, etc. New viruses or viruses with altered characteristics, including host range, were not detected in field-exposed or greenhouse-inoculated plants. The studies do not preclude any of the virus-virus interactions searched for, but do indicate that such interactions are rare events and that the risks of their occurrence may not be expanded by the fact that one of the genes is a transgene.

**Herbicide resistance: NewLeaf™ Roundup Ready**

Roundup (glyphosate) is a broad-spectrum herbicide used for the post-emergence control of annual and perennial weeds in major crop systems. It is rapidly degraded by naturally occurring soil microbes and is not a threat to groundwater or surface water. Nor will it accumulate in the food chain. It promotes environmentally favourable tillage methods, which allow farmers to protect soil from erosion and degradation. Several glyphosate-resistance crop species have been protected including cotton, maize and wheat. Glyphosate operates by inhibiting the action of the plant enzyme 5-enoyl-pyruvalshikimate-3-phosphate synthase (EPSPS) involved in the shikimic acid pathway which is responsible for the production of aromatic amino acids. Tolerance to glyphosate in GM crops is generated by expressing an EPSPS gene isolated from a strain of *Agrobacterium tumefaciens*. The gene encodes an EPSPS enzyme with reduced sensitivity to glyphosate compared with the plant enzyme. The enzyme therefore functions in essential aromatic amino acid biosynthesis in the presence of glyphosate. Glyphosate resistant potato has not reached the marketplace but, as with other glyphosate-resistant crops, could be expected to deliver more effective and precise weed control.

**Quality Traits: NewLeaf™ Anti-bruise Potatoes**

Blackspot bruise occurs on physical impact or following damage to tubers and can cause major losses to the commercial potato processor when producing chips (crisps) and French fries. Mechanical damage initiates enzymic browning and symptoms include production of black, brown and red pigments. Reduced bruise damage will help to minimize crop rejection and waste in processing lines due to automatic discarding of blackened fries and chips (crisps). Bruise resistance is a trait important to growers and processors alike. The reaction leading to pigment production is catalyzed by the enzyme polyphenol oxidase (PPO) which converts monophenols to *o*-diphenols and *o*-dihydroxyphenols to *o*-quinones. PPO activity has been reduced in potato by the company Keygene in the Netherlands which down-regulated PPO gene expression. Bachem et al., (1994) showed that these transgenics were less susceptible to bruising and De Both *et al.* (1996) showed in field trials that about 6% of the GM lines produced a bruise index of less than 10%. Some lines, with no detectable PPO transcript or enzyme activity, apparently show no enzymic browning. Recently, Coetzer et al. (2001) reported success using a tomato PPO gene in potato (co-suppression of PPO transcription).

The production of NewLeaf™ 'anti-blackspot' bruise potatoes has also involved down-regulation of a tuber-expressed PPO gene using a tuber-specific promoter. Lines of Russet Burbank that showed > 70% reduction in PPO activity were again almost devoid of any bruise symptoms under field conditions. The construct used has been optimized to increase the frequency of lines showing >70% reduction in PPO activity. This involved cloning a full length, tuber-specific PPO gene and testing three proprietary tuber-specific promoters in the cultivar Ranger Russet. All optimized constructs produced complete elimination of PPO and a corresponding commercial level of blackspot bruise resistance over three generation of replicated field trials. At the time when NatureMark®s business was closed, commercial line selection was in progress using transgenics with combined black-spot bruise and glyphosate resistance.

**NewLeaf™ Ultra**

This line is aimed at providing growers, processors and consumers with high yielding varieties with minimal chemical, agronomic and processing inputs. To this end a 'quadra' gene construct was made to deliver resistance to CPB, PLRV, PVY and tolerance to glyphosate. Expression of all four genes has been obtained in cvs. Russet Burbank and Atlantic and field level efficacies of the target traits demonstrated. This is the first known success in stacking four genes and obtaining desired performance in the field. Conceptually, the Ultra lines could be developed further to introduce resistance to Verticillium and improved quality traits such as elevated dry matter content and bruise resistance.

**Benefit Estimates from Potato Agricultural Biotechnology**

The National center for Food and Agricultural Policy (NCFAP) in the US published a report in January 2001 providing data on estimated benefits from commercializing GM crops. Data indicates that in 1996 only 1% of the US potato acreage was covered by GM varieties but by 2000 this had increased to between 2 and 3%. The figure reflect combined adoption of NewLeaf™, NewLeaf™ Plus (introduced in 1999) and NewLeaf™ Y (also introduced in 1999). The report suggests that factors which contributed to low adoption rates include the need to control other pests in addition to CPB (limiting potential savings on chemical insect control), the development of a very effective conventional insecticide (imidacloprid) and, of course, lack of adoption of GM potato by large industry end-users due to consumer concerns. However, the report does indicate that by using NewLeaf™ Plus growers could save considerably in reduced losses due to net necrosis and reduced insecticide costs. Including a 'technology fee' payable NatureMark® of $46 per acre for NewLeaf™ Plus, trials indicate an average saving of $85 per acre in 1998 and $134 per acre in 1999.

## Current and Future Potential for GM Potato

The following text provides examples of where future trends in GM potato production may lie, taking into account the potential for enhancing quality parameters, nutritional value and non-food uses. The examples selected are not meant to be comprehensive but illustrative.

### Antinutritional and Nutritional Compounds

The USDA-ARS has developed transgenic lines with reduced glycoalkaloid content by down-regulating the expression of a gene encoding solanidine UDP-glucosyltransferase. Glycoalkaloids are natural compounds which can be harmful to humans and animals when consumed in high concentration. Transgenic tubers show up to a 40% reduction in glycoalkaloid levels in field trials. This provides opportunities to rescue advanced breeding selections with excellent commercial traits but which were previously discarded due to unacceptable glycoalkaloid levels. Further benefits would accrue from reduced glycoalkaloids in potato starch wastes, as a high residue content renders the wastes unsuitable for use as fodder.

The production of transgenic crops containing protein with improved amino acid composition should be of benefit to humans as well as to monogastric animals (pig, poultry etc.) unable to synthesize all of the amino acids needed to sustain life. The potato is the most important non-cereal food crop in the world. However, it contains limited amounts of the essential amino acids lysine, tryptophan, methionine, and cysteine. Improvements in the nutritional value of food crops such as potato are especially important for people subsisting on a vegetarian diet in which the main source of protein comes from seeds, grains, tubers, etc., which contain limiting amount of essential amino acids. Chakarborty et. al., (2000) reported improvements in the nutritive value of transgenic potato through the expression of a non-allergenic seed albumin gene (*AmA1*) from *Amaranthus hypochondriacus*. As a donor gene, the *AmA1* gene has several advantages for genetic transformation experiments. First the seed protein has a well-balanced amino acid composition, making it nutritionally superior to other proteins recommended by the WHO. Second, the purified protein has no known allergenic properties. Finally, the protein is controlled by a single gene, which facilitates integration into other species. The team showed a five- to ten- fold increase in *AmA1* transcript levels in tubers of transgenic lines using the tuber-specific granule bound starch synthase (GBSS) promoter compared with the 35S CaMV promoter. Transgenic lines contained a significant two- to- four-fold increase in lysine, methionine, cysteine, and tyrosine content in their protein amino acids. Data collected for two consecutive years corresponded to an increase in most essential amino acids. Grain amaranth is used in many foods throughout the world and amaranth forage has been used for centuries as an important component of the human diet throughout the tropics. The authors presented these facts as evidence of the non-allergenic nature of amaranth.

Fructans, or fructose-oligosaccharides, consist of short chains of fructose molecules. Inulin is a mixture of linear fructose-polymers with different chain-length and a glucose molecule at each C-2 end. In over 30,000 plants (e.g. chicory, onion, asparagus, artichoke) inulin serves as a storage carbohydrate. Compounds such as inulin reduce the energy density of food and are used to enrich food with dietary fibre or to replace sugar and fat. When fructans are consumed, the undigested portion is reported to support growth of 'friendly' bacteria, such as Bifidobacteria and Lactobacillus species. Other benefits noted include increased production of beneficial short-chain fatty acids such as butyrate, increased absorption of calcium and magnesium and improved elimination of toxic compounds. Hellwege et al., have developed transgenic potato tubers which synthesize the full spectrum of inulin molecules naturally occurring in globe artichoke (*Cynara scolymus*). High molecular weight inulins have been produced by expressing the sucrose:sucrose 1-fructosyl transferase and the fructan:fructan 1-fructosylhydrolase genes from globe artichoke. Inulin made up 5% of the dry weight of the transgenic tuber. This approach has the potential to enhance the value of staple foods such as potato with compounds giving additional benefits.

## Food Processing and Industrial Uses

Starch is the primary storage compound in tubers and starchy foods are the world's most abundant staples. It is the most important source of calories in the animal and human diet and provides a starter material for the preparation of more than 500 different commercial products. The physical properties of starch vary with plant source but there are considerable opportunities to generate novel starches for use in food and non-food market sectors. Genetic engineering has already generated novel potato starch, including high amylopectin starch (with no apparent yield penalty) through the down-regulation of the granule bound starch synthase gene which controls amylose synthesis. High amylose starch is also in great demand by the starch industry for its unique functional properties, but very few high amylose crops are available. Scwall et al. (2000) showed that concurrent down-regulation of two starch branching enzymes, A and B, in potato tubers modifies both starch gain morphology and composition and produces a significant increase in amylose content.

Stark et al. (1992) increased the starch content of tubers by expressing an *E. coli* glgC16 gene which encodes for the enzyme ADP glucose pyrophosphorylase. The corresponding potato enzyme resides in the starch granule and plays a key role in starch biosynthesis. The *E. coli* enzymes is not regulated by the same fine control mechanisms which operate on the endogenous potato enzyme and is therefore able to increase the production of ADP glucose which becomes incorporated into the growing starch granule. Tuber starch content can be increased by up to 25% in some glgC16 expressing lines but the response appears to be genotype dependent. These high starch potatoes also accumulate lower levels of reducing sugars (glucose and fructose) in stored tubers which is highly relevant to the requirement of the processing sector. The processing industry requires low reducing sugar levels in tubers as these sugars are primarily responsible for non-enzymic browning through a typical Maillard reaction which occurs at the temperature required to generate potato chips (crisps) and French fries.

Ideally, the industry would like to store tubers at low temperature (ca. 4°C) to minimize sprout growth and eliminate the need to use chemicals to suppress the sprouting process. However, low temperatures induce glucose and fructose accumulation. Success in minimizing sugar accumulation using transgenic approaches have come from the use of the glgC16 gene and from modifying the expression of genes in pathways of primary carbohydrate metabolism, e.g., by minimizing the conversion of sucrose to glucose and fructose by expressing invertase inhibitor protein.

**Pharmaceuticals Uses**

Tacket *et al.* (2000) reported a new approach for delivering vaccine antigens using inexpensive plant-based oral vaccines generated in potato. Norwalk virus capsid protein (NVCP) assembled into virus-like particles, was used as a test antigen to determine immune response in healthy adults eating GM potato containing NVCP. Overall, 19 out of 20 volunteers developed an immune response of some kind. Similarly, Chong and Langride (2000) demonstrated expression of bioactive antimicrobial human lactoferrin in potato plants. This was the first report of synthesis of full length biologically active hLF in edible plants. Expression was significant (up to 0.1% of total soluble protein) and antimicrobial activity against four different human pathogenic bacterial strains was detected in extracts of tuber tissues.

At the time of writing this chapter, livestock and related industries in the UK are undergoing torrid times due to a severe outbreak of foot and mouth disease. Implementation of an expensive vaccination programme has been hotly debated. It is of some interest, therefore, that Carrillo et al. (2001) demonstrated the induction of a virus-specific antibody response to foot and mouth disease virus using the structural protein VP1 expressed in transgenic potato plants. The group previously reported the oral and parental immunogenicity of the structural protein VP1 of foot and mouth disease virus (FMDV) expressed in different transgenic plants. Their recent report indicates that transgenic potatoes containing the VP1 gene cloned under the regulatory activity of either a single or a double copy of the 35S CaMV promoter, represents a viable strategy for increasing the level of VP1 gene expression. Furthermore, immunized animals presented a FMDV VP1 specific antibody response and showed protection against the experimental challenge. These results clearly show the potential of using plants as antigen expression systems.

## Revised Legislation on GM Crops

The prospects for further commercial release of GM potatoes in North America, and of any of Europe, clearly await significant consumer acceptance before consistent, large volume end-user markets emerge. The factors which will determine acceptance or otherwise are complex and beyond the scope of this chapter. However, on February 2001, the European Parliament finally adopted a joint text concerning the deliberate release into the environment of genetically modified organisms. The text, which revises the European Directive 90/220/EEC (which itself lays down regulations for commercial release of GM crops into the environment), seeks to increase the transparency and efficiency of the decision-making process on GM crops and products in the European Union. This in itself may improve public confidence. The revised directive aims to promote a harmonization of risk assessment, and to introduce clear labelling requirements for all GMOs placed on the market. There are proposals to introduce mandatory monitoring for GM products and mandate a time limitation (renewable) of ten years, maximum, for first-time consent. There are plans to introduce compulsory monitoring of GM crops after they have been placed on the market, to provide for a common methodology to assess the risks associated with their release, and to include a mechanism to allow modification, suspension or termination of release when new information on risks become available.

Other important components of the evolving legislation include: a gradual elimination of antibiotic resistance markers in commercial GMOs by the end of 2004, and by 2008 for release into the environment for experimental purposes; a plan to bring forward a legislative proposal on environmental liability before the end of 2001, also covering damage resulting from GMOs; public registers of GMOs released into the environment for experimental purposes; introduction of general rules on traceability and labelling of GMOs and products derived from them; mandatory monitoring after GMOs are placed on the market; mandatory consultation of the public concerning both experimental and commercial releases; the application of the precautionary principle when implementing the Directive; the opportunity for consulting Ethics Committee(s) on issues of general nature.

## CONCLUSIONS

The commercial future of transgenic potatoes will clearly depend on end user 'pull' and the demand for distinct added value that these crops might provide. Such added value will be derived from reduced production and processing costs without yield penalties, improved quality and nutritional value, novel uses (diversification) and/or proven environmental benefits. It is clear that the benefits must be transparent to consumers who have increasing expectations with regard to food safety. The demands of growers and consumers will vary with geographical location and quality of life and a key challenge will be to provide proven biotechnological advances, technologies and products to the developing world at low, or zero costs, whilst maintaining profitability (and altruism) within the industrial sector.

Due to research and development costs this will be sustainable only if GM crops find their niche in global marketplaces. At present, Europe is a desert for GM crops and market resistance is evident in North America. NatureMark® products have figured significantly in this chapter since they have been the only company to place GM potato crops on the market. The demise of NatureMark® due to market forces may be seen as an opportunity by others to fill the gap. The number of mergers between plant biotechnology companies has been significant in recent years which potentially creates the 'monopolies' that many campaigners against GM crops would not wish to see. However, in the current climate the development of GM crops and products derived from them will require the 'patient money' that only large multinationals can provide whilst satisfying shareholder demands through other components of their business.

Due to the costs involved, opportunities for public sector research establishments to deliver GM crops to the market without strategic alliances with companies will be minimal to say the least. These are the realities; the industrial sector needs to make profit and the public sector needs to work closely with the industrial sector. However, this must not jeopardize independence, particularly where risk assessment is involved. The implementation of stringent risk assessment and risk management strategies together with labelling and traceability has become an important issue with the public. The recent revision of European legislation goes some way down this road.

Scientifically, one can predict an increased use in risk assessment of micro-arrays for large scale analysis of gene expression and proteomics and metabolic profiling for more detailed analysis of substantial equivalencies. Combined use of these approaches will also make major contributions to our understanding of metabolic networks and signal transduction mechanisms which govern important plant and crop traits. The explosion of sequence information delivered through genomics programmes will enable expression analysis on entire potato transcriptomes in the very near future, opening up immense research opportunities and challenges in information technology. One can predict the development of more efficient transformation systems, advances in transgene stacking to regulate several traits simultaneously (or traits which are under polygenic control) and to provide durable, multigene resistances to pests and diseases.

There is already a move towards the use of transformation vectors which are 'minimal' in the use of DNA sequences not required for the transformation event and which do not contain antibiotic resistance marker genes. The technologies are also available to sequence the regions bordering inserted sequences in the target plant. This would give information on potential disruption to important gene elements and might assist the targeting of metabolite analysis in a risk assessment exercise. However, it should be emphasized that the risk assessment procedures currently applied to GM crops are already far more rigorous than for any crops generated by traditional or mutation breeding.

# Chapter 9

# TRANSGENIC TOMATO

Tomatoes originated in the Andean region of South America under extremely variable climatic conditions. Wild relatives of tomato grow from sea level to subalpine elevations with some ecotypes adapted to flooded conditions and others to extreme drought. Domestication of tomato led to led to its cultivation as a crop on all continents and trials have been selected to promote abundant production of fruit. Selective breeding from the narrow genetic base of domesticated tomato as well as the introduction of exotic germplasm from the numerous wild relatives of tomato have developed tomato plants producing high-quality fruit for fresh consumption as well as for processed, prepared and stored products valued at approximately US$5 billion annually. Advances in agricultural biotechnology recently have provided the opportunity to expand the genetic resources available for tomato improvement.

The goals of tomato genetic engineering have been to protect the tomato crop from environment and biological assaults, and to improve the quality of tomato fruit in order to deliver greater value in processed tomato products or more healthful and attractive fresh fruit. Tomato fruit are a significant source of nutrition for substantial portions of the world's human population because this vegetable and concentrated processed products. Tomatoes are rich sources of vitamins, especially ascorbic acid and β-carotene, and antioxidants such as lycopene. A single small tomato is sufficient to supply about a quarter of the vitamins A and C recommended for humans to consume daily.

Most of the nutritional components in tomato fruit are stabilized by the acid pH of the fruit tissue and many of the human nutrients are conserved during the relatively short and mild processing used in preparation of most tomato food products. Tomatoes are grown in industrial quantities in many temperate locations, but the stability of the concentrated processed product has made it possible to transport tomato products widely and to prolong the storage of tomato products. Tomato was one of the first plants to be transformed by *Agrobacterium tumefaciens* and regenerated into fertile, productive plants. The success of early work to obtain transgenic plants allowed for the first commercial release of a transgenic food products, the Flavr Savr tomato, with extended shelf life of the ripe fruit.

The transformation of a large number of tomato varieties has been reported, suggesting that essentially and variety is amenable to genetic transformation. For example, fresh market varieties (Moneymaker, Better Boy) greenhouse varieties (Ailsa Craig), small-fruited fresh varieties (VFNT Cherry) and processing varieties (UC82b) as well as wild tomato relatives *L. chilense*, *L. peruvanium* and *L. hirsutum* have all been transformed in academic and commercial research laboratories. Most of the successful transformation protocols for tomato utilize *Agrobacteria* to deliver transgenes to the hypocotyl sections of newly germinated seedlings, but biolistic approaches also has not been utilized. The success of the floral dip methods used in *Arabidopsis* has not been reported for tomato.

Antibiotic resistance of transformed tissues is frequently the preferred method of selection of transgenic tissues, because of its historical success. However, public concerns abut the contents of genetically modified food products will undoubtedly lead to the utilization of new selection methods, including positive selection for growth on selective media. Because fruit are the economically significant crop from tomato plants, many transgenic modifications have targeted the fruit ripening processes to develop products that better withstand harvest, handling, transportation and storage practices utilized in commercial distribution. To reduce processing costs and effectively increase processing yield, transgenic fruit have been developed with increased solids content.

To provide novel value-added products, tomato fruit have also been engineered to produce increased components of nutritional value and to produce pharmaceutical compounds. To enhance production efficiency and yield, tomato plants have been

**Table 9.1. Transgenic tomato modifications**

| *Trait* | *Gene* | *Regulation* | *Expression* |
|---|---|---|---|
| **Fruit ripening** | | | |
| Ethylene reduction | Bacterial ACC deaminase | Constitutive | Expression |
| Ethylene reduction | Phage SAMase | Fruit specific | Expression |
| Ethylene reduction | Tomato ACC synthase | Constitutive | Antisense suppression |
| Ethylene reduction | Tomato ACC synthase | Constitutive | Sense suppression |
| Ethylene reduction | Tomato ACC oxidase | Constitutive | Antisense suppression |
| Fruit softening | Tomato fruit PG | Constitutive | Antisense suppression |
| Fruit softening | Tomato fruit PME | Constitutive | Antisense suppression |
| Fruit softening | Tomato fruit PG and PE | Constitutive | Antisense suppression |
| Fruit softening | Tomato fruit expansin | Constitutive | Sense suppression |
| Fruit abscission | Tomato fruit Cel1 and Cel2 | Constitutive | Antisense suppression |
| **Fruit composition** | | | |
| Sucrose accumulation, solid content | Tomato fruit invertase | Constitutive | Antisense suppression |
| Solid content | Bacterial *ipt* | Constitutive | Expression |
| Starch accumulation | *Arabidopsis sucrose synthase* | Constitutive | Expression |
| Fatty acid and flavour content | Yeast Δ9 desaturase | Constitutive | Expression |
| Colour | Tomato phytoene synthase | Constitutive | Antisense suppression |
| Parthenocarpic | Bacterial tryptophan monoxygenase | Constitutive | Expression |
| **Seeds** | | | |
| Increased dormancy | Tomato NCED | Constitutive | Expression |
| Decreased dormancy | *Arabidopsis abi-1* | Constitutive | Expression |

Continue...

| | | | |
|---|---|---|---|
| **Pathogen and pest resistance and tolerance** | | | |
| TMV | TMV N | Constitutive | Expression |
| CMV | Cucumber mosaic virus CP | Constitutive | Expression |
| TSWV | Tomato spotted wilt virus N | Constitutive | Expression |
| PhMV | Physalis mottle tymovirus CP | Constitutive | Expression |
| *Pseudomonas syringae pv tomato* | Tomato *Pto* | Constitutive | Expression |
| *Xanthomonas campestris pv. Vesicatoria* | Pepper *Bs2* | Constitutive | Expression |
| *Cladosporium fulvum* | Tomato *Cf9* | Constitutive | Expression |
| *Verticillium dahliae* | Tomato chitinase | Constitutive | Expression |
| *Fusarium oxysporum f.sp. lycopersici* | Tobacco chitinase and β1,3-glucanase | Constitutive | Expression |
| *Trichoderma hamatum* | Rubber tree hevein | Constitutive | Expression |
| *Xanthomonas campestris* pv. vesicatoria | Tomato *LeETR4* | Constitutive | Expression |
| *Sclerotinia sclerotiorum* | *Collybia velutipes* oxalae decarboxylase | Constitutive | Expression |
| *Botrytis cinerea* | Pear fruit PGIP | Constitutive | Expression |
| *Phytophthora infestans* | Grape reserveratrol | Constitutive | Expression |
| *Manduca Sexta* | Tomato prosystemin | Insect induced | Expression |
| Insect resistance | Bt toxins | Constitutive | Expression |
| Insect resistance | *Bt cry1Ac* | Constitutive | Expression |
| Nematode resistance | Rice cystatin *Oc-1* | Constitutive | Expression |

Continue...

| | | | |
|---|---|---|---|
| Root knot nematode, aphid, viral resistance | Tomato *Mi* | Root-specific | Expression |
| **Plant defense responses** | | | |
| Extracellular responses | *Agrobacterium ipt* | Constitutive | Expression |
| **Environmental stresses** | | | |
| Salt stress | Yeast *HAL2* | Constitutive | Expression |
| Drought | *Arabidopsis ABI-1* | Constitutive | Expression |
| Chilling and oxidative stress sensitivity | Tomato catalase | Constitutive | Antisense suppression |
| Heavy metal tolerance | Bacterial ACC deaminase | Root-specific or stress induced | Expression |
| **Herbicide resistance** | | | |
| Thiazopyr resistance | Rabbit liver esterase | Constitutive | Expression |
| Quinclorac resistance | Tomato ACC synthase | Constitutive | Antisense suppression |
| Fenthion (insecticide) sensitivity | Tomato *prf* | Constitutive | Antisense suppression |
| **Metabolic modifications** | | | |
| Increased sucrose unloading | Sucrose phosphate synthase | Root-specific or fruit-specific | Expression |
| **Foliage colouration** | | | |
| Increased anthcyanin | *Antirrhinum del* | Constitutive | Expression |
| **Floral patterns** | | | |
| Indeterminate flowering | Tomato *agamous* | Constitutive | Expression |
| Precocious termination | Tomato *agamous* | Constitutive | Antisense suppression |

engineered for resistance to herbicides, extreme temperatures and pathogens by the transgenic expression of foreign genes not accessible by classical breeding methods. A comprehensive listing of transgenic tomato modifications that have successfully altered aspects to plant growth, morphology and cultivation are summarized in Table 9.1

## MODIFICATIONS TARGETING FRUIT

Ripening is a genetically regulated developmental process that triggers metabolic changes enhancing the flavour, texture and aroma of fruit but that simultaneously initiates fruit senescence and deterioration. A major goal of tomato genetic engineering has been to manipulate the ripening process in order to delay fruit senescence and deterioration while retaining the beneficial metabolic attributes of the ripening process. Because of the importance of this developmental process, several approaches have been used to manipulate tomato fruit ripening.

The most general approach has been to modify the expression of regulators of suites of genes that control fruit development or ripening and over-ripening. Ethylene gas is produced by ripening tomato fruit and is the natural hormonal regulator of the ripening process. The amount of ethylene produced has been specifically modified in several lines of transgenic plants as a means to regulate the ripening process. A more specific approach to the regulation of aspects of ripening has been to modify the expression of genes whose produces encode enzymes or proteins that are instrumental in a targeted component of the process of ripening. In this regard, modification of the expression of various cell wall hydrolases has been attempted several times to suppress cell wall disassembly and fruit softening while allowing other aspects of ripening to proceed normally.

### Regulation of Ripening and Senescence

Tomato is a climacteric fruit and ripening is naturally regulated by ethylene produced by a the fruit at the onset of ripening. Two transgenic approaches have been used to reduce endogenous ethylene production in ripening fruit in order to delay the onset and rate of fruit ripening. The pathway of ethylene biosynthesis is now well known and the final two steps in the pathway, conversion of S-adenosyl methionine (SAM) to 1-aminocyclopropane-1-carboxylic acid (ACC) and its oxidization to ethylene have been targeted for modification in transgenic plants. One approach has been to metabolize either SAM or ACC to an inactive product and the second approach has been to specifically suppress the expression of the two ethylene biosynthetic enzymes required to catalyze the final steps in the pathway.

Both approaches resulted in fruit with significantly reduced ethylene content and greatly delayed ripening. To metabolize the ethylene precursors, SAM and ACC, microbial genes have been expressed in tomato. Metabolic inactivation of SAM was carried out by transgenic expression of bacteriophage T3 S-adenosylmethionine hydrolase (SAMase) gene in tomato. The SAMase enzyme converts S-adenosyl methionine, to methylthioadenosine and homoserine rather than ACC, and serves as a means to divert SAM from the ethylene biosynthetic pathway. The transgenic tomato plants were engineered to express SAMase only in ripening fruit after the breaker stage by linking the T3 SAMase coding sequence to the E8 promoter, a tomato fruit-specific and ripening-regulated promoter. Consequently, the effects of reduced ethylene were observed only in ripening fruit and these fruit exhibited delayed ripening and enhanced firmness.

An alternative strategy was to express a microbial gene encoding ACC deaminase, an enzyme that converts ACC to a-ketoglutarate in transgenic tomato. The ACC deaminase enzyme was identified in a strain of *Pseudomonas* that utilized ACC as a nitrogen source, and the corresponding gene was introduced into tomato to divert ACC from the ethylene biosynthetic pathway. The transgenic fruit exhibited reduced ethylene production and the fruit ripened at a slower rate and remained firm. No effects were observed in the vegetative tissues of these transgenic plants and, surprisingly, fruit ripening was delayed only in fruit detached form the plant but not in fruit allowed to ripen on the plant. Interestingly, transgenic tomato lines expressing a bacterial ACC deaminase with the root-specific *RolB* promoter or the pathogen inducible tobacco promoter (*PRB-1b*) were able to grow in the presence of heavy metals such as Cd, Co, Cu, Ni, Pb and Zn.

Transgenic tomato plants with reduced ethylene levels also were developed by suppressed expression of endogenous genes encoding the ultimate and penultimate steps in the biosynthetic pathway. Constitutive expression of a tomato fruit-specific ACC synthase gene in its sense orientation resulted in two phenotypically different groups of plants, those over-expressing the ACC synthase gene, and those in which ACC synthase gene expression was reduced by co-suppression of the endogenous gene.

The transgenic lines with reduced ACC synthase gene expression exhibited reduced ethylene production and reduced ripening. Another transgenic approach to reduce ethylene in ripening fruit relied on expression of antisense genes to suppress the expression of the endogenous ACC oxidase or ACC synthase genes. This approach was first explored as a means to deduce the function of a tomato gene (*pTOM13*) by antisense suppression of its expression. Fruit from these plants produced considerably reduced amounts of ethylene and ripened more slowly. Subsequently, *pTOM13* was shown to encode the ethylene-forming enzyme of tomato, ACC oxidase. This initial finding was later expanded to demonstrate that constitutive expression of the tomato antisense ACC oxidase gene caused delayed fruit ripening and delayed leaf senescence by 10 to 14 days.

A cDNA encoding apple fruit ACC oxidase has also been expressed in its antisense orientation in tomato, resulting in greater than 95% reduction in the endogenous tomato ACC oxidase mRNA accumulation, a reduction in the ethylene production, and delayed ripening. Theologis and colleagues also demonstrated that constitutive expression of an antisense ACC synthase gene significantly reduced endogenous ethylene production and delayed fruit ripening. Using these transgenic plants, the requirement for ethylene to initiate and maintain the progression of ripening and senescence was demonstrated. Altering the plants' perception of ethylene is another approach that has been taken to modify the role of ethylene in transgenic tomato plants.

The primary ethylene receptor is enclosed by the family of *ETR* genes, which in tomato consists of five members (*LeETR1*, *LeETR2*, *Nr*, *LeETR4* and *LeETR5*) that collectively control ethylene sensitivity throughout the plant. These genes encode histidine kinase sensors homologous to two-component signaling proteins found in bacterial and in *Arabidopsis*. *LeETR4*, *Nr* and *LeETR5* are expressed in ripening fruit and their expression is stimulated by ethylene, suggesting that the *ETR* genes are important in fruit for both the competence to respond to ethylene and the specific issue responses to ethylene. Expression of a mutant form of *Nr* in tomato renders the plants insensitive to ethylene. Fruit ripening was also delayed in transgenic tomato plants with antisense suppressed expression of *Nr*, although in this case suppression of *Nr* is compensated, at least in part, by an increase in *LeEXP4* expression.

Paradoxically, suppression of *LeETR4* expression in transgenic tomato results in increased ethylene sensitivity and premature flower senescence and more rapid fruit ripening. Ethylene regulation of *ETR* gene expression is complex, but future transgenic lines of tomato with specifically designed expression or suppression of multiple members of the *ETR* family may provide the basis to produce plants and tissues that exhibit precise responses to both endogenous and exogenous ethylene.

## Fruit Texture during Ripening

### *Endo-Polygalacturonases*

Fruit ripening is accompanied by disassembly of several cell wall polymers, including pectin and hemicellulose, which are primarily responsible for ripening-associated changes in fruit texture. Extensive studies on ripening-associated pectin disassembly and the expression of the endo-polygalacturonase (PG) gene family, suggested that tomato fruit texture could be modified by transgenic modification of PG gene expression. The expression of both antisense and sense constructs of the tomato fruit PG catalytic subunit (PG2) gene resulted in greater than 95% reduction in PG activity. Fruit with reduced expression of PG were analyzed for alterations in the expression of other cell wall hydrolases and none were detected. Fruit with reduced PG activity provided the basis for testing the significance of this PG during softening and ripening as well as the basis for the commercial introduction of fresh and processed tomato fruit whose texture was modified by this genetic modification.

Analysis of cell wall polymers of these fruit demonstrated that diminished PG expression contributed to reduced depolymerization of the chelator-solubilized pectins and increased viscosity of processed tomato products but did not reduce fruit softening. The effect of antisense suppression of a single fruit PG on fruit softening may be partially offset by expression of other tomato PG genes in ripening fruit. Interestingly, transgenic plants with reduced expression of fruit PG did not exhibit changes in leaf abscission, suggesting that PGs involved in abscission are distinct from those that participate in fruit ripening. The tomato fruit PG gene has also been inactivated by transposition and stabilization of a maize transposon, DS, within the PG gene.

Suppression of the non-catalytic β subunit of the PG1 isozyme complex in transgenic tomato plants by expression of an antisense gene construct also reduced pectin metabolism during fruit ripening. Specifically, the reduced expression of the PG1 β subunit, a regulatory subunit, reduced cell wall pectin solubilization and depolymerization, suggesting that the dynamics of pectin associations and structure in the cell wall are determined by several factors, perhaps some acting cooperatively. Several other fruit characteristics have been measured in tomato fruit with suppressed PG gene expression.

Transgenic tomato fruit were evaluated for sensory characteristics and their colour and flavour out performed a similar variety that was heterozygous for the *rin* (ripening inhibited), locus, a variety that had been bred for long shelf life. The tomatine content of transgenic fruit was unaffected by antisense suppression of PG. Furthermore, PG antisense fruit generally had improved integrity and were less susceptible to cracking and pathogen attack specifically at the over-ripe stage. However, the susceptibility of PG suppressed transgenic tomato fruit to *Colletotrichum gloeosporioides* was not measurably different than in wild-type fruit.

### *Pectin methylesterases*

Because methylation of pectins affects their structural properties in the celi wall as well as their susceptibility to pectinases, expression of pectin methylesterases (PMEs) has been altered to modify pectin metabolism of tomato fruit. Suppression of the expression of a single PME in tomato by the transgenic introduction of a truncated sense tomato PME gene resulted in significantly higher molecular weight pectins isolated from the fruit cell walls. Processed tomato products made from the PME suppressed transgenic lines also exhibited increased serum viscosity and reduced serum separation. Analysis of the PME mRNA in the transgenic plants suggested that the reduction in the endogenous PME mRNA resulted from interference by the transgenic mRNA with post-transcriptional processing of the endogenous PME mRNA. Transgenic expression of the sequence encoding 71 amino acids of tomato fruit PG linked to the tomato PME sequence under control of the 35S CaMV promoter resulted in suppression of both PG and PE simultaneously.

### *Expansins*

Expansins are cell wall proteins that have been proposed to participate in disruption of hydrogen bonding between hemicellulose and cellulose polymers at the surface of the cellulose microfibril. These polymeric associations are particularly important for cell expansion and other developmental events in which cell wall disassembly occurs, such as tissue softening during fruit ripening. Several expansin genes are expressed during tomato fruit development and ripening. The over-expression and suppression of one expansin gene. *LeExp1*, was examined in tomato plants expression a sense full-length or truncated *LeExp1* gene. Transgenic fruit from plants over-expression *LeExp1* were significantly less firm at the mature green and breaker stages and analysis of the cell wall material demonstrated that precocious depolymerization of the hemicellulose structure of the wall correlated with constitutive over-expression of *LeExp1*. However, expression in tomato of a cucumber hypocotyl expansin, *CsExp1*, did not result in phenotypic alterations of transgenic tomato fruit, suggesting that divergent expansin proteins may have distinct functions or substrates *in vivo*. Suppression of *LeExp1* in transgenic tomato resulted in increased fruit firmness, especially at the early (e.g. Breaker) stage of ripening. Surprisingly, in the *LeExp1* suppressed fruit, polyuronide depolymerization but not hemicellulose depolymerization was reduced in the later stages of ripening in these fruit.

### *β1, 4 endo-glucanases*

Because hemicellulose, and xyloglucans specifically, are disassembled in ripening fruit substantial research has focused on endo-β-1, 4-glucanases as a class of enzymes because they have the capacity to cleave the β-1,4-glucan linkages of xyloglucan. At least two endo-β-1,4-glucanases are expressed in ripening tomato fruit, and one of the, *Cel1* is also expressed in fruit abscission zones. Analysis of expression patterns demonstrated that the mRNA corresponding to a second endo-β-1,4-glucanase, *Cel2*, is more abundant in ripening fruit. However, the expression of both *Cel1* and *Cel2* during ripening suggested that the two endo-β-1,4-glucanases could act synergistically on their substrates in the softening cell wall. Transgenic plants in which *Cel1* expression was suppressed by an antisense gene construct produced fruit that softened normally, and abscission was partially reduced.

Transgenic plants engineered for suppression of the *Cel2* gene also exhibited no changes in fruit ripening or softening but also were altered in abscission zones, requiring greater force for the abscission zone breakage.

### *Galactosidases*

A recent report indicated that antisense suppression of galactosidase gene expression in ripening tomato fruit reduced by approximately 40%. At least four of seven tomato fruit β-galactosidases are expressed during ripening, and the release of galactosyl residues is the most dynamic cell wall change during ripening. As breakdown of galactose-containing polymers in the fruit cell wall is abundant during ripening, reduction of galactosidases, perhaps in concert with other cell wall hydrolases and expansin proteins may provide the basis to regulate the softening process of ripening tomato fruit.

## Fruit Composition

Soluble solids content of tomato fruit is a major determinant of fruit quality, particularly for processing tomatoes, and the soluble sugars. Thus, approaches to altering the composition of ripe fruit using transgenic strategies have focused on carbohydrate composition. One of the first attempts in tomato was to suppress the expression of acid invertase in ripe fruit in order to increase sucrose levels. Other transgenic tomato lines have been designed to alter the ratio of the monosaccharides, glucose and fructose, in the fruit.

### *Acid invertase and sucrose synthase*

Although tomatoes transport sucrose in the phloem, tomato fruit typically have very low levels of sucrose and approximately equal ratios of the hexose sugars, glucose and fructose. Interestingly some wild relatives of tomato accumulate primarily sucrose in their fruit and these fruit have very high levels of total soluble sugars. Introgression of the locus controlling sucrose accumulation (*sucr*) from the wild relative, *L. chemielewskii*, resulted in smaller fruit with increased soluble sugar levels. Because the *sucr* locus from *L. chemielewskii* was determined to encode an inactive allele of acid invertase, it was reasoned that the same trait could be produced by transgenic suppressio of invertase expression. Constitutive expression of an antisense gene encoding tomato soluble acid invertase resulted in tomato fruit with an increased concentration of sucrose and decreased concentrations of the hexoses, fructose and glucose. Fruit from the sucrose accumulating transgenic plants were approximately 30% smaller, presumably due to the osmotic effects of sucrose accumulation as compared to the hexose-accumulating non-transgenic control plants.

Many of the characteristics of the transgenic plants with reduced invertase expression were similar to sucrose accumulating lines of tomato that had been derived by introgression *sucr* locus from the *L. chemielewskii*. However, transgenic plants engineered using the E8 promoter to suppress invertase gene expression only in ripening fruit, remained hexose accumulators. This suggested that expression only in ripening fruit, remained hexose accumulators. This suggested that expression of the invertase gene regulates sucrose to hexose conversion early in fruit development, before the developmental timing of expression specified by the *E8* promoter.

Sucrose synthase has also been a target for genetic modification with the goal of enhancing sink 'strength' by increasing the capacity to metabolize imported sucrose. However, the fruit-specific antisense suppressio of sucrose synthase did not produce any changes in the accumulation of starch or sugars in the fruit tissues even thought he transient increase in sucrose synthase expression normally observed in early in fruit development was suppressed. Similar transgenic plants with suppressed expression of a fruit-specific sucrose synthase also exhibited no change in hexose or starch accumulation. However, fruit set (e.g. number of fruit) on these plants was diminished and the sucrose unloading capacity of young fruit was significantly reduced.

Additional transgenic strategies have been employed to specifically alter source-sink relations in tomato by ectopic expression of sucrose phosphate synthase. When the *Zea maize* sucrose phosphate synthase gene regulated by the rubisco promoter was expressed in transgenic tomato foliar tissue sucrose partitioning was increased and this reduced limitations of photosynthesis. Sucrose unloading in fruit also was increased by the over-expression of sucrose phosphate synthase.

### *Hexokinase and fructokinases*

Over-expression of hexokinase in transgenic tomato demonstrated the regulatory role of this enzyme in photosynthetic tissues, particularly affecting senescence. However, in fruit from plants over-expressing an *Arabidopsis* hexokinase gene the quantity of starch in young fruit and of hexose in ripe fruit was reduced. Because fructose is almost twice as sweet as glucose, modification of fructokinase expression in ripening tomato fruit has been attempted in an effort to increase the ratio of fructose an glucose and, thus, enhance fruit 'sweetness'.

Two genes encoding fructokinase are expressed in tomato fruit, *Frk*1 and *Frk*2. *Frk*2 expression is co-related with periods of starch accumulation, and *Frk*1 is expressed ubiquitously, although less abundantly. The potentially complex regulation of both the expression and activity of each of these potentially complex regulation of both the expression and activity of each of these fructokinase isoforms may confound attempts to increase fructose concentrations in fruit from transgenic plants with antisense genes for either or both fructokinases.

**Fruit Flavors and Aromas**

Fruit flavor is a complex trait determined by the mixture and balance of sugars, acids and a large number of aldehyde, ketone and alcohol volatiles. Lipoxygenases are key enzymes in fatty acid metabolism, producing the hexanal aldehyde and alcohols that contribute substantially to the volatile tomato aromas and flavors. At least three lipoxygenase genes (*LOX*) are expressed during tomato fruit ripening. The expression of the *LOX* genes in ripening tomato fruit is regulated by both ethylene and developmental factors. In order to modify the C6 aldehyde and alcohol composition of tomatoes, antisense transgenic tomatoes were developed using a conserved region of lipoxygenases to potentially suppress expression of all LOX activities.

In ripe fruit of these transgenic plants, the expression of *tomlox*A and *tomlox*B, but not *tomlox*C, was substantially reduced. However, no changes in the flavor and aroma constituents resulted, suggesting that regulation of their biosynthesis may be complex or that *tomlox*C expression was sufficient for these pathways to proceed unimpeded. Another approach to altering the flavour composition of tomato fruit was engineered by the expression of a yeast Δ9 desaturase gene, resulting in changes in the fatty acid content of fruit. Concentrations of palmitoleic acid, 9,12-hexadienoic acid, and linoleic acid increased and concentrations of palmitic acid and stearic and were reduced. Concentrations of flavor compounds derived from these fatty acids were changed and resistance to powdery mildew was enhanced. Another approach to changing the composition of flavour aldehydes in tomato fruit was taken by altering the expression of alcohol dehydrogenase 2 (*adhII*). The amount of Z-3-hexanol decreased and the amount of 3-methylbutanal increased in plants with reduced expression of *adh*II. Fruit from plants with increased expression of *adhII* had a more intense 'ripe fruit' flavour and had increased amounts of hexanol and Z-3-hexanol.

**Fruit Colour and Vitamin A**

Tomato fruit colour is primarily determined by the concentration of the red carotenoid, lycopene, and its precursors. Because these carotenoid pigments are the source of provitamin A as well as the other major nutritional antioxidants in tomato, their levels have significant consequences for both the appearance and nutritional value of fresh an processed tomatoes. Modification of the carotenoid composition of tomato fruit has been achieved by transgenic modification of the activity of a key enzyme in the carotenoid biosynthetic pathway, phytoene synthase that converts phytoene to lycopene. Suppression of phytoene synthase gene (*PsyI*) expression significantly reduced carotene and xanthophyll as well as abscisic acid (ABA) in ripe fruit compared to control fruit. The resulting yellow fruit were very similar to naturally occurring phytoene synthase mutant fruit.

Plants constitutively over-expressing *PsyI* are dwarf, apparently because gibberellin biosynthesis is reduced, and have less chlorophyll in their leaves. A second phytoene synthase gene (*Psy2*) also is expressed in ripening fruit but apparently is not involved in carotenoid synthesis in this tissue. Recently, in an effort to increase the vitamin A content of tomato, the expression of a bacterial phytoene desaturase in tomato has been reported. The *Erwinia uredovora crtI* gene was expressed constitutively in tomato plastids. The total carotene composition of these plants was unaltered although the proportion of β-carotene increased moderately. The lycopene produced in the transgenic plants was cyclized by the induction of two endogenous lycopene cyclases to increase the β-carotene content about two-fold in these orange fruit. Unlike retinol or vitamin A, β-carotene or provitamin A is non-toxic and can be stored by after human ingestion. Thus, this transgenic modification provides an effective nutritional supplement of about of about 40% of the recommended daily consumption of vitamin A in a single fruit.

## Other Fruit Characteristics

Fruit development is promoted by auxin produced by seeds developing within the ovary walls. Parthenocarpic fruit or fruit which develop in the absence of ovule fertilization, have the obvious advantage of being seedless and in tomato also have been reported to have elevated soluble solids content. Parthenocarpic tomato fruit were generated by the transgenic expression of a microbial gene (*iaaM*) that encodes a tryptophan monoxygenase used in synthesizing an auxin precursor. The *iaaM* gene from *Pseudomonas savastanoi* was linked to *DefH9* promoter from *Anthirrhinum majus* that specifies expression in placental and ovule tissues, resulting in auxin production targeted to the ovary tissues destined for fruit development.

As in other plants, the additional IAA in the developing fruit promoted fruit formation in the absence of pollination or ovule fertilization. The composition, size and abundance of the fruit from these plants are indistinguishable from control fruit. Similar strategies have been used to alter the levels of other plant hormones in developing tomato fruit. The fruit-specific expression of a bacterial *ipt* gene specifying elevated cytokinin biosynthesis in ovary tissues of tomato resulted in increased total and soluble solids. The fruit from these plants had islands of green pericarp tissue and excess cytokinin from fruit was exported to the leaves causing increased accumulation of genes, such as PR-1 and chitinase, induced by cytokinin. Transgenic approaches to alter fruit size and shape have not been reported. However, the identification of the QTL locus, *fw2.2,* in the introgressed population between *L. esculentum* and *L. pimpinellifolium* may suggest candidate genes that could be introduced transgenically into tomato varieties which will influence tomato fruit size.

## Modification Targeting Seeds and Germination

Modification of the seed germination process may be possible as gene expression patterns during germination have been recently reported. The expression of mannanase in the seed endosperm and a PG expressed at the time of radicle emergence were candidate marker genes for early germination. However, the expression of both of these genes was most abundant in the later stages of germination and not early enough to serve as a marker of the initial stages of germination. The expression of a second mannanase (*LeMAN2*) gene in the emerging micropylar endosperm appears to be more directly related to the earlier stages of germination.

The identification of expansins in germinating seeds has provided the clearest opportunity for marking the germination process in individual seeds. *LeEXP4* is expressed inn the micropylar region and *LeEXP8* is expressed in the radicle tip and their expression is precisely correlated with germination. These seed germination-specific promoter elements may provide the means to monitor germination of individual seeds or to enhance or accelerate seed germination as a means to enhance stand establishment. Seed specific expression of a GUS reporter gene in tomato also has been noted using portions of the maize *Sh1* promoter. Seed dormancy can be promoted by increased amounts of ABA. One of the outcomes of over-expressing a tomato 9-cis-epoxycarotenoid dioxygenase (NCED) in tomato was increased seed dormancy as well as increased stomatal conductance resulting from elevated ABA levels. Some plants from this same population exhibited co-suppression of the endogenous NCED gene and had a wilted phenotype.

## Modification Targeting Biotic and Abiotic Stress Tolerance

### Pathogen Resistance

#### *Viral resistance*

Tomato plants engineered for increased resistance to viral pathogens have been developed by expression of specific components of viral genes in transgenic plants. For example, transgenic expression of the TMV coat protein (CP) in tomato increased resistance to viral infection without altering the nutritional and biochemical constituents of the fruit. Expression of a truncated form of the replicase gene of the tomato yellow leaf curl geminivirus (TYLCV) in tomato provided resistance to this virus but not to another geminivirus, tomato leaf curl virus (ToLCV-Au). Expression of the cucumber mosaic virus CP in transgenic tomato provided durable resistance to CMV under field conditions, and expression of CP from more than subgroup of CMV in other transgenic tomato lines has provided broader resistance to strains of CMV even in epidemic conditions in the field.

Expression of CMV satellite RNA in tomato also increased CMV resistance. Resistance to the tomato spotted wilt virus (TSWV) has been achieved by expression of the nucleocapsid protein (N) in transgenic tomato lines. Partial resistance and delayed symptom development to Physalis mottle tymovirus (PhMV) has been obtained by expressing the coat protein of this virus in tomato. Endogenous plant genes that confer virus resistance also have been employed to engineer for virus resistance in tomato. For example, expression of the tobacco race-specific *N* gene in tomato also conferred resistance to the appropriate strain of TMV. This demonstrated that the tobacco *N* gene functions in tomato and can be used as a source of genetic resistance.

### Bacterial resistance

Resistance to *Pseudomonas syringae* pv tomato is specified by the products of a host-specific plant resistance gene (*Pto*) and a bacterial avirulence gene (*avrPto*). The sufficiency of the *Pto* gene to confer resistance to the bacterial strains producing *avrPto* was demonstrated by the transgenic expression of *Pto* in tomato. *Pto* encodes a kinase and autophosphorylation is required for the resistance hypersensitive response, including the *avrPto* specific oxidative burst. *Prf*, a gene related to *Pto* in the resistance cluster, when expressed in transgenic plants, confers sensitivity to the insecticide fenthion. Resistance to bacterial speck disease is observed in transgenic tomato plants in which the pepper *Bs2* resistance gene is expressed. This disease is caused by *Xanthomonas campestris* pv. *vesicatoria* strains expressing the *avrBs2* gene.

### Fungal resistance

Resistance to fungal pathogens has been improved in tomato by the transgenic expression of race-specific resistance genes, targeting initially particular pathogens, and by the over-expression of proteins that augment or activate defense responses to a broader group of fungal pathogens. The approach of expressing resistance factors that recognize specific microbial avirulence factors has demonstrated that these factors are sufficient for gene-for-gene resistance, and several examples have shown that the expression of these factors also can provide some resistance against other pathogens.

The over-expression of recognition factors and genes that activate defense responses against pathogens has incrementally improved resistance to fungal pathogens. Expression of the *Cf9* gene, the extracellular and membrane anchored leucine-rich repeat resistance factor against *Cladosporium fulvum* race 9 in tomato, demonstrated that this gene is sufficient to confer resistance in an otherwise susceptible variety of tomato. Furthermore, the expression of *Cf9* has allowed details of interaction between *C. fulvum* and resistant tomato varieties to be examined. Expression of the bacterial gene, *avr9*, in a line of tomato to be crossed with *Cf9* tomato, demonstrated that expression of the *avr* gene in plant cells triggers the hypersensitive response. A $K^+$ channel of the transgenic tobacco leaf guard cells is apparently involved in the recognition of *avr9* by the host and the interaction involves at least one phosphorylation step.

### Insect and nematode resistance

Expression of *Bacillus thuringiensis* (*Bt*) toxins has been used in many plant species to control insect pests on transgenic plants. The expression in transgenic tomato of the synthetic *Bt cry1Ac* gene coding for an insecticidal crystal protein (ICP) provided a high level of resistance to the larval stage of the fruit borer, *Helicoverpa armigera*. Systemin, a systemic signal molecule that triggers endogenous defense mechanisms in tomato, is produced by proteolysis is a precursor protein, prosystemin. The transgenic expression of prosystemin in tomato yielded tomato plants with improved resistance to the tomato hornworm pest *Manduca sexta*.

As a result of the constitutive presence of systemin, these plants produced tow proteinase inhibitor proteins and a polyphenol oxidase that participate in endogenous tomato defense mechanisms. Mutants of the transgenic tomato plants constitutively expressing systemin have been used to identify a group of genes that are required for wound and system induced defense gene expression. Thus, engineering systemin expression has provided additional understanding about the regulation and function of components of defense pathways elicited in tomato vegetative tissues by this mobile signal molecule. The expression of cystatin, a cysteine protease inhibitor or papain inhibitor, is induced in plants over-expressing prosystemin and pest-induced expression of cystatin provides an inducible mechanism for insect resistance.

Expression of the rice cystatin Oc-I in tomato improved the resistance of roots to the nematode pathogens *Meloidogyne incognita* and *Globodera pallida*, reducing the size of the female nematodes and frequency with which they fed on the transgenic roots. Expression of the PHI-*ipt* cytokinin biosynthesis gene in tobacco increases the toxicity of foliar tissues to *Manduca sexta* and *Myzus persicae*, but expression of the same gene in tomato did not reduce the feeding capacity of these pests. In tomato, a well-known source of nematode resistance (*Mi* gene) was originally identified in the wild tomato relative, *L. peruvianum*, and has since been introgressed into most commercial tomato cultivars. The recent cloning of the tomato *Mi* gene has provided an opportunity to improve resistance to root knot nematode species (*Meloidogyne* sp.) as well as aphid species (*Macrosiphum euphorbiae*).

Root-specific expression of the *Mi* gene has been possible using *A. rhizogenes* transformation and identification of the functional roles of some portions of the protein have been clarified by transgenic expression of modified forms of *Mi*. A homologue of *Mi, Sw-5*, provides resistance topovirus species, suggesting that viral and nematode resistance may share portions of a defense signalling pathway. The transgenic expression of *Mi, Sw-5* or other related variants that include the domains responsible for the virus and nematode resistance, provides an unprecedented opportunity to engineer effective resistance in tomato against viral, nematode and insect pests simultaneously.

### *General defense*

Transgenic expression of genes encoding proteins involved in plant defense against many pathogens or in the regulation of defense responses has been used to elucidate the significance of specific components of plant defenses or as a strategy to reduce susceptibility to group of pathogens. Expression of the *Agrobacterium ipt* gene resulted in increased cytokinin synthesis in tomato and the constitutive activation of extracellular defense responses such as *PR-1* and acidic chitinase gene expression that normally occur in response to elicitors, including *Fusarium oxysporum* cell wall components. Cell cultures established from transgenic tomato lines expressing either cytokinin or auxin biosynthetic genes from *A. tumefaciens* and were assessed for the effects on *F. oxysporum* growth.

Cell cultures from a transgenic susceptible tomato variety became more resistant when cytokinin was increased as a result of *ipt* expression, and a transgenic resistant variety expressing the auxin biosynthesis gene reduced *F. oxysporum* growth. Expression of *A. tumefaciens ipt* in tomato also increased resistance to *Manduca sexta* and *Myzus persicae*. Over-expression of *L. chilense* chitinase in *L. esculentum* improved foliar resistance to *Verticillium dahliae*. Simultaneously expressing tobacco chitinase and β1,3-glucanase in tomato reduced symptoms of *Fusarium oxysporum* f.sp. *lycopersici* by about 50%, suggesting that these proteins can act synergistically to defend against a fungal pathogen.

Expression of the precursor of the chitin binding protein, hevein, in tomato retarded the growth of *Trichoderma hamatum*, but did not eliminate infections. The hevein prepeptide was not cleaved efficiently in the transgenic plants to form mature hevein. Antisense suppression of an anionic peroxidase in tomato was attempted in order to define the role of this enzyme in plant defenses. Although peroxidase expression was reduced, no influence on suberin and cell wall phenolics was ascertained, but the plants exhibited some resistance to immature stages of the insects, *Helicoverpa zea* and *Manduca sexta*.

Over expression of the tobacco anionic peroxidase in tomato increased lignin composition, but did not provide increased resistance to pathogens. Over-expression of the tomato basic peroxidase, *tpx1*, resulted in increased lignin in the transgenic plants but no effect on pathogen susceptibility was reported although some changes in stress responses were observed. As ethylene synthesis and perception are signaling components for the activation of many defense responses, alterations in ethylene and its recognition had significant effects on pathogen susceptibility. However, the outcomes of ethylene modifications vary depending on the tissue, stage of development, and pathogen.

Many pathogens induce ethylene synthesis and the role of ethylene receptors in *Xanthomonas campestris* pv. *vesicatoria* infections of tomato has been explored by examining the *X. campestris* susceptibility of transgenic lines expressing the ethylene receptors, *NR* or *LeETR4*. Both lines of transgenic plants respond to the pathogen with reduced necrosis and reduced ethylene sensitivity because *NR*, like *LeETR4* is a negative regulator of ethylene sensitivity. In non-transgenic tomato, the induction of ethylene receptors by pathogens may reduce ethylene sensitivity and, hence the expansion of necrosis.

Inactivating products or eliminating functions that pathogens use in virulence have been used to engineer resistance in tomato.

An improvement in resistance to *Sclerotinia sclerotiorum* has been accomplished by expressing an oxalate decarboxylase form *Collybia velutipes* in tomato. As *S. sclerotiorum* requires oxalate for infection, the removal of oxalate by this transgenically expressed enzyme represents a novel approach for limiting the growth of the pathogen on plant tissues. Transgenic expression in tomato of a plant inhibitor (PGIP) of a fungal cell wall hydrolase (PG) has proven to be a means of reducing the spread of *Botrytis cinerea* infections on leaves and fruit. *B. cinerea* uses PG, along with several other cell wall hydrolases, to break down plant tissues during establishment of decomposing lesions. Transgenic synthesis of the phytoalexin reserveratrol in tomato enhanced resistance to *Phytophthora infestans* but not to *B. cinerea* or *Alternaria solani*.

## Salt, Water and Temperature Stress

Tolerance of growth in high salt conditions is an attribute that has been amenable to transgenic manipulation in tomato. Expression of the yeast *HAL2* gene in tomato allows hypocotyl growth and root formation on high salt media. Plants expressing the yeast *HAL1* gene probably accommodate high salt conditions by increasing the intracellular $K^+$ concentration. Over expressing the *A. thaliana* vacuolar $Na^+/H^+$ antiport in tomato allows growth in up to 200 mH NaCl. Drought tolerance can be influenced by the ABA concentration in tomato.

Engineering plants with an *Arabidopsis* wild type *ABI-1* or a mutant *abi-1* gene has demonstrated that transgenic modifications affecting ABA concentration modulate responses to drought. Plants expressing *abi-1* are wilty and have reduced seed dormancy. These plants are affected as well in their responses to wounding. Decreased tolerance to chilling and oxidative stress was confirmed by the antisense suppression of catalase that resulted in an increase in the $H_2O_2$ concentration in the plants. Transgenic tomato plants expressing an *E. coli* glutathione reductase were equally sensitive to chilling as wild-type plants, suggesting that ascorbate peroxidase pathway in the chloroplast is not limiting for the chilling sensitivity of photosynthesis in tomato.

## Herbicide Tolerance

Resistance to the herbicide thiazopyr has been engineered by the transgenic expression of a rabbit liver esterase. Seedlings deactivated the herbicide in proportion to the expression of the transgene. Transgenic tomato plants with suppressed expression of the ACC synthase gene do not display deleterious effect of the herbicide quinclorac. Treatment of wild-type tomato with ethylene inhibitors similarly alleviates the effect of this herbicide.

## Modifications Targeting Vegetative Tissues and Flowers

Increased pigmentation of tomato vegetative tissue has been achieved by the transgenic expression of the delila (*del*) gene from *Antirrhinum*. The anthocyanin pigment is regulated by the expression of the *del* gene and linking the maize AC transposable element has resulted in variegated plants, suggesting that the *del* gene acts cell autonomously. The result of expressing *del* in *Arabidopsis* and in tobacco differ from the results of expressing the gene in tomato. The responses of tomato plants to light exposure have been modified by the over-expression of avena phytochrome A (*PhyA*).

The response of seedlings to 'end of the day' exposure to light occurred largely at low Pfr/P (3–61%) for the wild-type and at high Pfr/P (61–87%) for transgenic seedlings. Tomato plants with reduced *ACC* oxidase expression because of the transgenic expression of *pTOM13* demonstrated reduced yellowing in response to long day length light exposure. Alterations in floral morphology and cell fate can be achieved by he over-expression or suppression of the tomato *AGAMOUS* homologue, *TAG1*. Intermediate flowering can be achieved by the over-expression of *TAG1* and precocious termination and differentiation by the antisense suppression of *TAG1* expression.

## Expression of Novel Proteins in Tomato

The expression of pharmaceutically relevant proteins in tomato has been pursued as a possible approach to facilitate the production of delivery of compounds that enhance human and animal health. The rabies glycoprotein, G-protein, was expressed in transgenic tomato. The protein was glycosylated uniquely probably because of its expression in tomato, but the expression of the rabies virus G-protein provided an opportunity to attempt to produce an orally accessible vaccine in plants. A potential oral vaccine to the respiratory syncytial virus (RSV) has been developed by expressing the RSV fusion (F) protein in ripening tomato fruit using the *E8* promoter. Durable serum and mucosal RSV-F antibodies were generated in mice who had consumed the ripe tomato fruit from these transgenic plants. Other pharmaceutical compounds, including human β-interferon have been reported to be produced in non-traditional varieties of tomato.

## Regulation of Transgenic Gene Expression in Tomato

As the technology for expression of transgenes in plants becomes more sophisticated, tissue specific, developmentally timed, and inducible expression of specific genes will be required. One way to ensure appropriate expression in through the choice of promoter regulating gene expression. Several promoters have been evaluated in tomato for the specificity of their expression. Fruit-specific expression has been important for genes whose products influence fruit performance or content.

The E8 promoter specifies ethylene-regulated fruit-specific expression, as does the E4 gene promoter. The ca. 2kb promoters for ACC oxidase and PG from apple also function in fruit to specify expression during ripening. The tomato fruit PG promoter (1.4kb) gene directs expression during fruit ripening. Expression specified by the PG promoter is confined to the outer pericarp tissue of the fruit and is not ethylene regulated. Fruit-specific expression also can be achieved using the tomato 2A11 promoter. The tomato lipoxygenase genes A and B (*tomloxA* and *B*) are expressed in the pre-ripening stages of fruit development and a third tomato lipoxygenase gene (*tomloxC*) is expressed in germinating seed. The promoter for *tomloxA* is able to direct gene expression in the outer pericarp from 5–20 days postanthesis. Gene expression is confined to tomato sink tissues and stimulated by sucrose when regulated by the potato ADP-glucose pyrophosphorylase promoter.

Expression in ripening tomato fruit also can be specified by the promoters from pepper capsanthin/capsorubin synthase and fibrillin genes, genes whose expression is ripening regulated in the non-climacteric pepper. The expression from these promoters is ethylene regulated in tomato but also responds to other stimuli such as drought. The expression of the ethylene inducible genes in tomato leaf abscission zones and adjacent petioles can be specified by the soybean cellulase promoter and by the promoter for the tomato abscission zone specific PG. The abscission zone PG promoter does not cause expression inducible by ethylene in fruit tissues. Expression of genes with a circadian expression pattern can be specified by the light harvest complex protein promoter.

**Table 9.2. Promoters used in transgenic tomato**

| *Specificity* | *Promoter* | *Regulation* |
|---|---|---|
| Fruit-specific | Tomato2A11 | |
| | Tomato PG2 | Outer pericarp |
| | Tomato E8 | Ethylene and ripening induced |
| | Tomato E4 | Ethylene induced |
| | Tomato LoxA | Outer pericarp |
| | Apple ACC oxidase | Ripening |
| | Apple PG | Ripening |
| | Pepper capsanthin/ capsorubin synthase | Drought and ethylene induced |
| | Pepper fibrillin induced | Drought and ethylene |
| | Tomato HMG2 | Coincides with lycopene synthesis |
| | Tomato RBCS | Locular specific in developing fruit |
| Sucrose sink tissues | Potato ADP-glucose Pyrophosphorylase | Sucrose regulated |
| Leaf abscission zones | Soybean cellulase | Ethylene induced |
| | Tomato abscission zone PG | |
| Flower-specific | Tomato LAP | Methyl jasmonate induced and induced in wounded leaves. |
| | Tomato LAT52 and LAT 59 | Pollen and late anther specific |
| Germinating seeds | Tomato LoxC | |
| | Tomato LeEXP8 | Micropylar region specific |
| Vegetative tissue | Maize Sh-1 | Not expressed in fruit |
| Circadian expression | Tomato Lhc | Clock controlled |
| Pathogen induced | Tomato prosystemin | Jasmonate induced, insect feeding induced |
| Drought induced | Tomato H1-S histone | ABA induced in all tissues |
| Metal regulated | Tomato metallothionein | Expressed more in leaves than in roots |

Gene expression in tomato flowers can be achieved using the promoter for the tomato leucine aminopeptidase (*LAP*). The promoter for systemin has been linked to the insecticidal proteins because the promoter is activated by wounding or insect attack. Gene expression from this promoter is induced by methyl jasmonate as well. Gene expression in response to ABA or drought in tomato can be specified by the H1-S histone protein promoter. Pollen-specific expression in tobacco has been reported. Development an wound regulated expression of the promoter for the tomato anionic peroxidase promoter has been reported in tobacco but not in tomato.

## CONCLUSIONS

With the exception of *Arabidopsis*, tomato is perhaps the best genetically characterized dicotyledonous plant. Because of its rich genetic history, the availability of numerous mutants and genetic history, the availability of numerous mutants and genetic stocks and its relatively easy transformation by *Agrobacterium*, tomato has been used as a model crop to test the effects of numerous transgenes. Although tomato was the first genetically engineered food to be commercialized with the release of the Flavr Savr tomato in the early part of the 1990s, today there are no commercial transgenic tomato cultivars.

The commercial production of transgenic corn, soybeans and cotton most likely reflects the relatively higher value of the major agronomic crops and lack of commercial research attention to the majority of horticultural crops, including tomato. Because tomatoes play a central role in diet of many cultures, this fruit provides a unique vehicle for the delivery of vitamins and other healthful constituents to the human diet. Transgenic approaches will provide the essential genetic tools to enhance or redirect metabolism towards constituents that provide new human health benefits and to provide a high quality product for consumers. As the agricultural biotechnology industry matures and public acceptance of genetically modified foods grows, we can anticipate that the extensive research base in tomato will lead to the commercial development of a large number of transgenic tomato cultivars.

# Chapter 10

# Transgenic Rice and Maize

A continued increase in food production will be essential in order to sustain an increasing world population. This must be achieved by the development, for example, of higher yielding varieties with improved nutritional quality and tolerance to biotic and abiotic stresses. Whilst conventional breeding will play a major role in increasing crop yield, it is clear that laboratory-based techniques, such as genetic transformation to introduce novel genes into crop plants, will be essential in completing existing breeding technologies. Since cloned genes are introduced into target cultivars during transformation, this eliminates the requirement for repeated sexual back-crossing to remove undesirable co-transferred into cereals, such as rice, by transformation is more extensive that the range of genes which could be introgressed using other somatic cell techniques such as interspecific somatic hybridization (involving protoplast fusion), form the genome pool of wild species. A decade ago, cereals were considered recalcitrant to transformation but since that time, their genetic transformation has become more efficient, as considerable research has been focused on these crops because of their agronomic status.

Globally, rice and maize are two of the major cereals, with rice providing the staple diet of more than one-third of the world's population. Consequently, it is not surprising that these plants have become prime targets for genetic manipulation involving the introduction of foreign DNA. Initially, the genetic transformation of cereals, such as rice and maize, relied upon the use of experimental systems centred mainly on the ability of protoplasts isolated from suitable tissues of a limited range of target cultivars to take up DNA and to produce callus from which fertile plants could be regenerated. More recently, major advances have been accomplished in the regeneration of fertile plants form a range of source tissues, providing an essential foundation for the generation of transgenic plants.

Interestingly, the transformation of rice and maize has progressed more rapidly than that of other cereals, which reflects the progress made in the tissue culture and regeneration of rice and maize. Thus, for several cultivars, the transformation technology is now routine, although procedures still need to be refined to maximize transformation of specific cultivars. The exploration of transformation technology has already resulted in transgenic maize and rice plants expressing agronomically useful characteristics, such as resistance to herbicides, insects, fungi and viruses, examples of which are discussed later in this chapter. The literature relating to maize and rice is now so extensive that it is impossible to include reference to all published work and experimental details in this chapter. Consequently, the citation of references is limited, where possible, to the most recent and relevant reports.

## Approaches to Transformation of Maize and Rice

Birch discussed the requirements for efficient transformation systems. Essentially, the transformation protocol must be cultivar-independent, technically simple and safe to operate. The target tissue must be regenerable, readily available and the time required in culture should be minimal to reduce both cost and somaclonal variation. Ideally, selection procedures must be efficient, at low copy number without incorporating unwanted vector sequences from outside the T-DNA, or encountering variable transgene expression, due to position of insertion of the transgene or multiple copy-induced transgene co-transformation and stable integration of multiple genes is essential.

Additionally, the possibility of removing reporter and selectable marker genes from transformed plants is desirable and may become mandatory in the future. Several experimental procedures are available to introduce genes into target plants, each approach having its own merits and limitations. Those methodologies which have been evaluated extensively for rice and maize, include DNA uptake into isolated protoplasts, the use of biolistics, electroporation, electrophoresis and silicon carbide whiskers and, more recently, *Agrobacterium*-mediated gene delivery.

**Protoplast-based Technologies**

During the late 1980s and early 1990s, emphasis was placed on the use of protoplasts isolated from embryogenic cell suspensions of cereals as recipients for foreign DNA. Such procedures have been reviewed by Mass et al. In general, embryogenic callus initiated from tissues, such as immature zygotic embryos, is transferred to a liquid medium for the production of rapidly growing, homogeneous cell suspensions containing a high proportion of densely cytoplasmic, actively dividing cells. Enzymatic digestion of the walls of such cells releases populations of protoplasts (naked cells), which are mixed with foreign DNA. Protoplasts take up this DNA through their plasma membrane when treated with chemicals such as polyethylene glycol (PEG) and/ or high voltage electrical pulses. Several difficulties are associated with this technique. Although about 50% of the protoplasts take up exogenous DNA, the final transformation frequently is low, and is usually in the order of 1 protoplast in 105 being stably transformed. Furthermore, the development of this technology relies upon efficient protoplast-to-plant regeneration systems, which are often genotype dependent.

The production of cell suspensions which relies totipotent protoplasts is, theoretically, simple. However in practice, it is laborious, time consuming, and riles heavily upon the intuition of the worker to identify and to select at an early stage of culture those cells which are most likely to produce cell suspensions suitable for protoplast isolation. In addition, aberrant plants are often regenerated from transformed protoplast-derived tissues, frequently with multiple copies of the foreign DNA associated with complex patterns of integration into the plant genome. Despite these limitations, protoplast transformation enabled production of the first transgenic rice and maize plants. Unfortunately, the transgenic plants regenerated in these early experiments were sterile. Subsequently, however, Golovkin et. al. introduced a mutant dihydrofolate reductase (DHFR) gene from mouse that confers methotrexate resistance into maize protoplasts using PEG and generated fertile plants, with cross-pollination experiments permitting segregation analysis of the transgene in seed generations. Interestingly, more recent studies have produced evidence that protoplast transformation can be used in maize to generate transformants with single-copy well-defined inserts of foreign DNA.

The latter authors introduced the *virD1*, *virD2*, and *virE2* genes from the Ti plasmid of *Agrobacterium tumefaciens* into protoplasts on a plasmid, which carried the genes of interest flanked by the *Agrobacterium* 25 bp T-DNA repeat sequences. The latter were also from the Ti plasmid. The presence of the *virE2* gene gave maximum transformation. Unfortunately, such experiments were stimulated by recent advances in *Agrobacterium* mediated transformation of cereals and by improved knowledge of the molecular biology of *Agrobacterium*-plant interactions, especially the relevance of virulence (*vir*) genes in T-DNA transfer. In other experiments, Tsugawa et al. reported the use of a synthetic polycationic amino polymer (polycation) for rice protoplast transformation, but this system requires further assessment. An extensive review of maize transformation using protoplasts is presented by Armstrong. Similarly, Tyagi et al. have summarized recent progress to 1997 in protoplast-mediated rice transformation for the production of herbicide-, fungal-and insect-resistant plants. It is likely that there could be a resurgence of interest in DNA uptake into isolated protoplasts for rice and maize transformation in view of the most recent achievements.

**Biolistic**

Biolistics, coined from the term 'biological ballistics', is also known as microprojectile bombardment, particle acceleration, particle bombardment or gene gunning. It involves the delivery of delivery of tungston or gold particles (usually gold because they are chemically inert) of a suitable size (0.4-1.2 μm) coated with DNA into plant cells. Since initial reports of this technology the equipment has been improved to exploit electrical discharge or helium pressure for more controlled and consistent particle acceleration. Fertile, transgenic maize was first generated by Gordon-Kamm et al. following microprojectile bombardment of embryogenic cell suspensions, while Christou et al. initially applied this procedure to rice to introduce DNA into immature embryo of indica and japonica cultivars. Currently, this technology is in routine use for cereal transformation in several laboratories worldwide, utilizing totipotent cells as targets for transformation.

As in the case of protoplast-mediated transformation, the literature reporting stably transformed rice and maize has been reviewed by Tyagi et al. and Armstrong respectively. The advantages of microprojectile bombardment are that there is no requirement for the development of labour-intensive, cultivar-specific protoplast-to-plant systems, and it bypasses any host specificity associated with *Agrobacterium*-mediated gene delivery. The system is both cultivar- and species-independent, simple to perform and transgenes can be introduced into any tissues of any plant genotype.

The commercial availability of biolistic equipment, such as the BioRad PDS 1000/He device, has facilitated the standardization of gene delivery parameters in different laboratories although other devices, such as the particle inflow gun or custom-built instruments, have been used to deliver genes into rice, maize and other cereals. Pareddy et al. discussed several aspects of maize transformation by microprojectile bombardment using 'helium blasting' and described the principle and design of two novel devices for rapid DNA delivery and/or aiming capabilities in the production of transgenic maize. Noteworthy is the fact that Sudhakar et al. have described the use of a portable, inexpensive helium-driven particle bombardment device, which is an improvement of the particle inflow concept and which operates without vacuum. Such an instrument, that gives transformation rates comparable to those of other more sophisticated devices, should facilitate the transfer and development of this technology to laboratories, which to date, have been unable to purchase more expensive instruments. Additionally, it may facilitate gene delivery to field-grown plants.

Important features relating to optimization of transformation by biolistics include the nature of the microprojectiles, the use of calcium chloride and spermidine to aid adherence of DNA to microprojectiles, the choice of DNA construct the target explants, together with the physical bombardment parameters. The latter include the flight distance of particles in the instrument, the helium pressure and vacuum conditions. The conditions of the target cells is crucial in maximizing transformation. For example, in the case of immature zygotic embryos, the donor plant growth conditions must be optimal, avoiding exposures to stress conditions such as pest or disease attack, or suboptimal watering, temperatures and humidity. Poor donor plant growth conditions affect explant physiology and the competence of cells for transformation.

The physical parameters and conditions influencing microprojectile bombardment are described by Christou and Southgate et al. Chen et al. have reported their protocol for consistent, large-scale production of transgenic rice plants using biolistics, while Hagio has provided suggestion for improving and maximizing transformation efficiency using this gene delivery technique.

### *Agrobacterium*-mediated Transformation

The molecular biology of *Agrobacterium*-mediated transformation of dicotyledons is now relatively well understood together with the possible mode of insertion of the DNA transferred from the Ti or Ri plasmid (the T-DNA) into the recipient plant genome. An essential feature of the process is that the *virD1* and *virD2* genes of the Ti or Ri plasmid encode for endonucleases specific to the defined ends (25 bp borders) of the T-DNA, which ensure controlled DNA transfer with minimal rearrangement on entry into the recipient plant genome. Foreign DNA inserted between the borders of the T-DNA is introduced into recipient plant cells. Gene delivery into dicotyledons by *Agrobacterium tumefaciens*, and to a lesser extent by *A. rhizogenes*, has been exploited extensively for several years with T-DNAs, disarmed by removal of their oncogenicity genes, being used to effect gene delivery. Such disarmed T-DNAs are either retained on the Ti plasmid, or removed from the Ti plasmid and inserted into a smaller plasmid.

The Ti plasmid, whilst deleted of its T-DNA, still retains its *vir* genes to effect T-DNA transfer from the smaller plasmid to recipient plant cells. Such *Agrobacteria* thus carry two plasmids, constituting the binary vector system. *Agrobacterium*-mediated gene delivery is simple, efficient when optimized, and transfers low copy number, intact transgenes. However, the procedure has been more difficult to apply to cereals and other monocotyledons, since they are not infected naturally by *Agrobacterium*. Although *Agrobacterium*-based procedures have been evaluated in several laboratories over a number of years, with early reports of T-DNA transfer into maize and rice, these early reports for *Agrobacterium*-mediated transformation of monocotyledons were viewed with scepticism. In a milestone publication, Hiei et al. provided unequivocal evidence for the production of hygromycin-resistant GUS-expressing plants of japonica rice cultivars, which stimulated the development of *Agrobacterium*-mediated transformation systems for other cereals, such as maize. Undoubtedly, several factors influence the efficiency of *Agrobacterium*-mediated transformation of rice and maize. Specifically, the use of actively-driving cells such as those in embryogenic scutellum-derived callus, exposure of bacterial cells to the phenolic wound signal molecule, acetosyringone, to activate *Agrobacterium vir* gene expression, the use of 'super-virulent' *Agrobacterium* strain and the development of 'super-binary' vectors combined with an appropriate selectable marker.

In their report, Hiei et al. evaluated a range of explants for transformation, together with *Agrobacterium* strains commonly used to transform dicotyledons, namely LBA4404 and the 'super-virulent' EHA101. Transformation was most efficient with scutellum-derived embryogenic callus (in common with the general observations for biolistics) of the japonica rice cultivar Koshihikari. Both EHA101, carrying the binary vector pIG121Hm, and LBA4404, with the vector pTOK233, transformed rice cells. LBA4404(p-TOK233) contained, on the binary vector, the *virB* and *virG* genes from the *A. tumefaciens* Ti plasmid Bo542, resulting in a 'super-binary' plasmid. This strain/vector combination was superior to other strains evaluated, due to the increased virulence resulting from the amplified expression of the vir genes. Up to 28.6% of scutellum-derived tissues inoculated with LBA4404(p-TOK233) produced transgenic plants. Subsequently, other workers have reported reproducible *Agrobacterium*-mediated transformation of the indica rice cultivars Basmati 385 and Basmati 370 by EHA101 (pIG121Hm), the japonica cultivar Radon and the indica cultivars IR72 and TCS10 by LBA4404 (pTOK233), and the commercially important USA javanica) cultivars Gulfmont and Jefferson with the same constructs and strains developed by Hiei et al. Zhang et al. also used LBA4404(pTOK233) to generate fertile transgenic plants of the indica cultivar Pusa Basmati 1, while Azhakanandam et al. confined the efficiency of the same construct to transform cultivars of japonica, indica and javanica subspecies of rice. All these reports confirmed transgene integration into the genome of recipient plants, generally at low copy number, although single copy T-DNA inserts were rare. Sequence analysis revealed that the borders of the T-DNA in transgenic rice were essentially identical to those in *Agrobacterium*-transformed dicotyledons. Transmission of transgenes through seed generations was Mendelian, as confirmed by the other reports. Collectively, these results confirmed the suitability of this experimental approach for the biotechnological improvement of rice.

Importantly, *Agrobacterium*-mediated transformation has been extended to other cereals, with the first report of maize transformation by Ishida et al. using immature embryos as target tissues. An interesting concept is that of Trick and Finer, who showed that exposing tissues to brief periods of ultrasonication in the presence of *Agrobacterium* results in a 100- to 1400-fold increase in transgene expression in tissues of several crop plants, including maize. Apparently, ultrasonication produces small and uniform fissures and channels between plant cells, facilitating access of *Agrobacterium* into the internal plant tissues. A similar approach has been reported by Zhang et al. resulting in the production of fertile maize plants transformed with an insecticidal protein (Bt) gene from *Bacillus thuringiensis*, the acoustic intensity and duration of treatment being important parameters in the procedure.

Recent developments in *Agrobacterium*-mediated cell transformation, including the production and analysis of various *Agrobacterium* strains vectors, are discussed by Tyagi et al. Hiei et al. and Komari et al. An interesting concept in vector development has been the use of procedures that combine the advantage of *Agrobacterium* transformation with the efficiency of biolistic delivery to cereals. Thus, Hansen and Chilton described a novel 'agrolistic' system in which the *virD1* and *virD2* genes from *A. tumefaciens*, that are required in the bacterium for excision of T-strands from the T-DNA of the Ti plasmid during gene transfer to plant cells, were placed under the control of the CaMV35S promoter. The *vir* genes were co-delivered into maize cells by bombardment with a target plasmid containing T-DNA border sequences flanking the gene of interest. *Vir* gene products *in planta* caused strand-specific nicking at the right T-DNA border sequence, similar to *virD1/virD2* catalyzed T-strand excision which normally occurs in *Agrobacterium*. Some inserts in transformed cells exhibited right border T-DNA junctions with plant DNA that corresponded precisely to the sequence expected for normal *Agrobacterium* T-DNA insertion events into the genome of dicotyledonous plant cells. It will be interesting to assess the applicability of this novel system for rice transformation.

## Microinjection, Silicon Carbide Whiskers and Electrical Procedures

The cost of reliable microinjection equipment combined with the difficulty of penetrating cell walls, and the need to avoid rupturing the vacuole allowing toxic vacuolar contents to enter the cytoplasm, have discouraged workers from applying this labour-intensive procedure to the transformation of cereals. In spite of these limitations, Leduc et al. reported the transient expression of the beta-glucuronidase (*gus*) gene and anthocyanin reporter genes following microinjection of excised maize zygotic embryos. In general, the real application of this technology is in assessments of gene expression following DNA injection into individual cells, since it is not a realistic procedure with which to generate large populations of transgenic plants.

An interesting use of microinjection has been the introduction of individual *Agrobacterium* cells into single meristematic cells of maize to effect transformation. Although young, immature maize embryos appear to lack competence for transformation, this approach proved that maize meristematic cells are in fact, competent for transformation by *Agrobacterium*, although the response is genotype dependent. The vortexing of embryogenic cell suspensions with DNA and silicon carbide fibres (whiskers), has been suggested as a simple, inexpensive procedure for the generation of transgenic plants. Using this approach, Nagatani et. al. obtained 533 transformants in rice per gram fresh weight of target cells. Whilst other workers have assessed this procedures, they have obtained evidence for only very low levels of transgene expression in treated cells of maize and have been unable to regenerate transgenic plants.

The attraction of such a simple procedure is that it would be easy to transfer to other laboratories. However, evidence is yet to been presented for the reproducibility of this procedures. Although the cell wall is often considered a barrier to DNA transfer, which may be overcome by removal or weakening of the wall by exposure to cellulytic and pectolytic enzymes, or by wounding, there are claims that electroporation (electropulsation) with high-voltage electrical pulses can deliver DNA to intact cells and tissues. In maize, this approach has been exploited to introduce the gus gene and genes for chloramphenicol acetyltransferase (*cat*) and phosphinothricin acetyltransferase (*bar*) into intact cells of black Mexican sweet maize, cell plasmolysis before electropulsation being essential in the transformation process. Other workers have also transferred genes into immature embryos and Type II (embryogenic) callus of the inbred maize line A188. The results of the latter authors were comparable to those of Pescitelli and Sukhapinda for transgene (*gus*) expression in Type II callus of the 'High-II' hybrid and the 'back-crossed B73' genotypes. However, in contrast to the report of D'Halluin et al. enzyme treatment of maize tissues was not required to facilitate DNA entry into recipient cells.

Immersion of target cells in the electroporation buffer may be sufficient to permit diffusion of DNA through cell walls in readiness for its passage through the plasma membrane during electroporation. Success has also been reported in transforming the elite Italian rice cultivars Lido, Cornaroli and Thaibonnet by electroporation of suspension cultured cells, the transgenic plants exhibiting, importantly, negligible genomic changes. Tissues electrophoresis has also been assessed for DNA delivery into cereals, DNA being believed to migrate into target tissues when the latter are exposed to a low voltage electrical field for several hours. Although there are reports of this procedure being used to introduce DNA into maize embryos, Southgate et al. were unable to obtain evidence for gene introduction into embryogenic maize tissues. Laser beam-mediated gene transfer to rice has also been described. Overall, whilst techniques such as microinjection, the use of silicon carbide whiskers, tissue electrophoresis and laser beams are available, they will probably continue to have limited application to cereal transformation.

## Target Tissues for Rice and Maize Transformation

Successful plant transformation necessitates efficient DNA delivery into recipient cells, which must be capable of rapid cell division and plant regeneration. Consequently, the choice of tissue for transformation is crucial. Transformation of protoplasts by electroporation or PEG treatment requires a population of homogeneous cells for enzyme treatment and the release of protoplasts. Cell suspensions, leaf mesophyll tissue and aleurone layer cells have been used as a source of protoplasts, but, in general , embryogenic cell suspensions are most suitable for protoplast-based cereal transformation. A long-term, non-regenerable suspension of the maize cultivar Black Mexican Sweetcorn (BMS) has been particularly useful in developing transformation procedures, in optimizing protocols and in assessments of DNA constructs.

As already indicated, a major problem with protoplast transformation has been the production of sterile or morphologically abnormal plants from transformed tissues, which may be related to the long periods (often several months) from initiation of the cultures to the time when the suspensions are homogeneous and the cells are in the correct developmental stage of release totipotent protoplasts. In this respect, embryogenic scutellum-derived tissues may be most suitable for the rapid generation of cultures for protoplast production. Initially, immature embryos were used to develop procedures for microprojectile bombardment of maize. Klein et al. conducted short-term (transient) gene expression studies with the maize cultivar A188, but failed to achieve stable transgene integration. Gordon-Kamm et al. generated the first fertile, transgenic maize plants following microprojectile bombardment of embryogenic cell suspensions. Later, bombardment of immature embryo scutella or callus resulted in the production of fertile transgenic maize plants.

Currently, immature embryos, immature embryo-derived callus and embryogenic cell suspensions are the main targets for microprojectile bombardment-mediated transformation of maize, although Zhong et al. reported a novel and reproducible transformation system using cultured shoot apices. Additional factors that influence transformation include explant pre-culture to initiate cell division prior to transformation, and/or treatment with a high osmoticum mediated to induce cell plasmolysis. The latter reduces cell damage and leakage of cytoplasm on impact of high-velocity DNA-carrying particles. As Songstad et al. stated, the advantage of bombarding scutellum-derived tissues is their ability to regenerate shoots rapidly, the latter exhibiting increased male and female fertility compared to shoots regenerated from long-term callus or cell suspensions. The earliest report of cultivar-independent method for biolistic-mediated transformation of rice was by Christou et al. who targeted immature zygotic embryos. Subsequently, Cao et al. bombarded embryogenic cell suspensions.

As in the case of maize, the targets currently used for routine transformation of rice include immature embryos, embryogenic callus and embryogenic cell suspensions, derived from immature or mature embryos. The review by Tyagi et al. gives further details describing the biolistic-mediated transformation of rice. In evaluating several tissues of rice for *Agrobacterium*-mediated transformation, Hiei et al. concluded that scutellum-derived embryogenic calli gave the highest transformation frequency. Suspension cultures were unsuitable for *Agrobacterium* transformation unless the cells were transferred from liquid to semi-solid culture medium for a period prior to bombardment.

Later, Aldemita and Hodges described the use of immature zygotic embryos for efficient *Agrobacterium*-mediated transformation, while Park et al. inoculated isolated shoot apices of rice to produce transgenic plants. The review papers of Hiei et al. and Tyagi et al. give comprehensive details of reports describing *Agrobacterium*-mediated transformation of rice. In extending *Agrobacterium*-mediated transformation to maize, Ishida et al. followed the procedures originally developed for rice, using co-cultivation of immature maize embryos with bacterial cells. Maize transformation using *Agrobacterium* is still in its early stages and most research has been limited to the cultivar A188. Less commonly used target cells for cereal transformation include microscope anther culture-derived tissues and leaf bases.

## Vectors for Rice and Maize Transformation

The development of cereal transformation technology has relied on the use of a limited number of constructs composed of constitutive promoters, reporter genes and selectable marker genes. More recently, efforts have focused on the identification of tissue-specific and developmentally regulated promoters, in combination with agronomically useful genes. Vectors for basic studies of transformation and the development of reliable protocols usually consist of a reporter gene and a selectable marker gene, each driven by a suitable promoter. Subsequently, vectors have also included genes for agronomically useful characteristics. Promoter sequences are essential in order to control gene expression in target plant cells. Since the constitutively expressed cauliflower mosaic virus (CaMV) 35S promoter has been used extensively and successfully in the transformation of dicotyledons, it follows that this promoter has also been evaluated for cereal transformation.

As already discussed, the 'super-binary' vector pTOK233 has been particularly useful in rice transformation by *A. tumefaciens*, the vector carrying the hygromycin phosphotransferase (*hpt*) and *gus* genes, both with the CaMV35S promoter. In general, the CaMV35S promoter, combined with the *gus* reporter gene and the *hpt* selectable marker gene, has been the focus of *Agrobacterium*- and protoplast-based transformation procedures, whereas a larger range of promoters have been utilized in microprojectile bombardment studies of both rice and maize. Overall, the 35S promoter has proved to be less effective in monocotyledons than in dicotyledons. Consequently, several alternative promoters have been tested for rice and maize transformation. The cereal-derived *act1*, *ubi1* and *emu* promoters are particularly effective, since naturally they drive gene expression to high levels in maize and rice. They have been assessed in studies of rice and/or maize m transformation, in association with reporter and selectable marker genes, including the *gus*, *bar*, *hpt* and neomycin phosphotransferase (*nptII*) genes. Compared to the 35S promoter, the *ubi1*, *act1* and *emu* promoters gave improved transient (short-term) expression following microprojectile bombardment of rice and maize.

**Table 10.1. Promoters used in cereal transformation studies**

| *Promoter* | *Construct* |
|---|---|
| CaMV35S-*adh1 intron1* | CaMV35S promoter, enhanced with the first intron from the maize alcohol dehydrogenase (adh) gene |
| *act1-act1 intron1* | Rice actin 1 promoter |
| *adh1-adh1 intron1* | Maize alcohol dehydrogenase promoter |
| *ubi1-ubi1 intron1* | Maize ubiquitin promoter |
| *emu* | Modified *adh1* promoter and first intron |

The introduction of introns associated with promoter sequences in genes constructed for cereal transformation has enhanced expression in transformed tissues. For example, Vain et al. compared a range of monocot and dicot intron-containing fragments inserted into the 5' untranslated leader between the promoter and transgene (*gus*) of interest. In this study, the maize *ubi1* intron gave a 71-fold enhancement of gene expression in maize. Interestingly, enhancement of gene expression was not limited to introns from monocotyledons, since the dicot *chsA* intron gave a 92-fold enhancement of gus expression in maize. However, as these authors emphasized, it does not necessarily follow that an intron which is effective in maize will be equally effective in other cereals.

**Table 10.2. Selectable marker genes for cereal transformation**

| *Gene* | *Enzyme encoded* | *Confers resistance to* |
|---|---|---|
| nptII | Neomycin phosphotransferase | Neomycin, kanamycin, geneticin/G418, paromomycin |
| hpt | Hygromycin phosphotransferase | Hygromycin B |
| Bar | Phosphinothricin acetyltransferase | Phosphinothricin (PPT), bialaphos, glufosinate ammonium, Basta |

Since major advances have been accomplished in recent years in our knowledge of rice genetics, this cereal is now considered a model for both *Agrobacterium*- and biolistic-mediated transformation of cereals. Consequently, research is more advanced in rice than in other monocotyledons. Current strategies are focused on the identification and development of tissue-specific promoters. Maize and rice transformation studies have involved several of these promoters, including tapetum-specific promoters (e.g. *Osg6B*) and endosperm-specific promoters (maize waxy gene, zein, *CM3*). The promoter most suitable for driving transgene expression can be confirmed by linking, at least initially, the promoter to a suitable reporter gene, followed by assessments of gene expression during short-term or longer periods (stable transformation).

A reporter gene should have no negative effects on plant metabolism, it product should be stable *in vivo* and the detection assay should be simple, low cost and sensitive. Any endogenous gene activity should be minimal in order not to mask reporter gene expression. The gus, anthocyanin and luciferase (*luc*) reporter genes have been used in rice and maize, the most commonly used being the *gus* gene, derived from *Escherichia coli*. Beta-glucuronidase, the enzyme encoded by the *gus* gene, catalyses the hydrolysis of a range of substrates, including X-Gluc (5-bromo-4-chloro-3-indolyl-β-D-glucuronic acid), the product of which forms an indigo dye within transformed plant cells. The disadvantage of this assay is that it is destructive. Nevertheless, the *gus* assay is an improvement on previous systems, such as the CAT assay, which requires radioactive reagents.

Other reporter genes used for assessments of both transient and stable transformation include the *nptII* gene whose expression can be assessed by ELISA, and the C1, B and R genes which regulate anthocyanin biosynthesis. Their expression results in red pigmentation in transformed cells. The luciferase gene from the firefly (*Photinus pyralis*), which catalyses the oxidation of D(-)-luciferin in the presence of ATP, leads to the accumulation of oxyluciferin and emission of yellow light, although visualization requires expensive equipment. A relatively new marker is the green fluorescent protein (gfp) from the jellyfish *Aequora victoria*, the assay of which is simple, requires no additional substrate and is non-destructive. Selectable marker genes permit the growth of only transformed cells on selective medium, usually by conferring resistance to antibiotics and herbicides.

The *nptII* gene, conferring resistance on plant cells to antibiotics such as kanamycin sulphate and geneticin (G418), was used in early studies of DNA uptake into cereal protoplasts and in *Agrobacterium*-mediated gene delivery. However, cells of cereals can be naturally resistant to kanamycin sulphate and this antibiotic also inhibits shoot regeneration from transformed tissues. Similarly, selection on a medium containing G418 results in poor organogenesis, although this antibiotic is less inhibitory than kanamycin to shoot regeneration. Those shoots which to regenerate in the presence of kanamycin or G418 are often albinos. The *hpt* gene, from *E. coli*, encoding for hygromycin resistance, is more effective for rice.

Additionally, the bar gene from *Streptomyces hygroscopicus* is both a selectable marker and introduces an agronomically useful characteristics, herbicide resistance, into transformed cells and regenerated plants. It is interesting to note that genomic DNA can also be used to effect transformation. Thus, Sawahel used total genomic DNA isolated from a hygromycin-resistant maize cell line to transform hygromycin-sensitive lines to antibiotic resistance. The merit of this procedure is that it circumvents the requirement for gene cloning and, consequently, could see application in the future for the introduction of agronomically useful genes into target cultivars. A possible limitation may be the size of DNA used, since this appears to influence transformation.

## Examples of Agronomically Useful Genes Introduced into Rice and Maize

Understanding the expression of simple reporter and selectable marker genes in transgenic plants is important in predicting the behaviour of agronomically useful genes introduced into crops. Consequently, it is imperative to assess gene expression in large seed populations. In this respect, Zhong et al. studied expression of the *bar*, potato proteinase inhibitor II and *uidA* (*gus*) genes, confirming their co-integration, co-inheritance and co-expression in 286 first generation (T1) plants and in 11,000 second seed generation (T2) maize plants. Similarly, Brettschneider et al. followed the inheritance and expression of transgenes to the fourth seed generation in several inbred lines and sexual hybrids of maize. In field studies, Oard et al. assessed expression of the bar gene, giving resistance to the herbicide glufosinate, in the commercial rice cultivars Gulfmont, IR72 and Koshihikari. They confirmed that the *bar* gene was effective in conferring field-level resistance to the herbicide in rice, although, importantly, variation amongst transgenic lines required traditional breeding selection procedures to identify plants with high levels of herbicide resistance. These workers also emphasized the need to generate several independent transgenic lines of each cultivar for transgene assessments.

Recent studies have reported the development of transgenic plants containing agronomically useful genes, in addition to those for herbicide resistance. Since insects cause substantial crop losses worldwide, it follows that engineering plants for insect resistance has, and will continue, to receive high priority. Stem-boring insects are common pests in maize and rice, and resistance against these insects has been achieved, primarily, by the introduction and expression of modified or synthetic versions of the *Bt* δ-endotoxin, a natural insecticidal toxin from *Bacillus thuringiensis*. For example, Wunn et al. and Cheng et al. introduced the *cry1A(b)* gene into rice cultivars, including the indica cultivar IR58, to confer resistance to yellow stem borer (*Scirpophaga incertulas*) and striped stem borer (*Chilo suppressalis*), while Alam et al. were the first to engineer a lowland deep water rice for stem-boring insects, expression of the *cry1A*(*b*) gene also inhibited feeding of the leaf-folding insects *Cnaphalo crocis* and *Marasmia patnalis* on transgenic rice.

Comparisons have been made of the expression of the *Bt cry1A(b)* gene driven by different promoters, including the constitutive 35S and Actin-1 promoters, with tissue-specific promoters from pith tissue and the pep-carboxylase (PEPC) promoter form chlorophyllous tissue of maize. The latter promoter gave high levels of transgene expression in leaves and stems of rice. Modified versions of the *cry1A(b)* gene have been used in rice transformation to give plants which induced 100% mortality in feeding yellow stem borers. Synthetic truncated genes, based on the *cry1A(b)*, gene have also been introduced into rice. The latter authors targeted low tillering aromatic rices which are particularly difficult to improve by conventional breeding because of loss of quality characteristics upon sexual hybridization.

The *cry1A(c)* gene has also been assessed in rice transformation for stem borer resistance, the *cry2A Bt* gene also conferred resistance to yellow stem borer and rice leaf folder insects in the indica rice Basmati 370 and M7. A recent example of the transformation of maize for insect resistance is that of Fearing et al., who introduced the *cry1A(b)* gene into six commercial cultivars and four backcross generations. They reported the highest concentration of insecticidal protein to be at anthesis in transformed plants. Other genes conferring insect resistance which have been evaluated in rice, include the Cowpea trypsin inhibitor (*CpTi*) gene, which increased resistance of transgenic rice to striped stem borer and the pink stem borer, and the snowdrop lectin (GNA) gene. The latter was directed against sap-sucking insects, such as the brown plant hopper, through the use of a rice sucrose synthase promoter to drive GNA expression in the phloem of transgenic plants. Nematodes cause severe crop losses in some areas, including rice cultivated in Africa.

In attempts to reduce nematode damage, a cysteine proteinase inhibitor (oryzacystatin-I Delta D86) gene was introduced into four elite African rice cultivars, resulting in a 55% reduction in egg production by the nematode *Meloidogyne incognita* in the roots of transgenic plants. Viral and fungal disease reduce crop productivity. The insertion of viral coat protein genes into transgenic plants is a well-established procedure for conferring viral resistance, this approach being exploited in rice for resistance to rice dwarf virus. Rice has also been engineered for resistance to sheath blight incited by the fungus *Rhizoctonia solani*. Thus, introduction of a 1.1 kb fragment of a rice chitinase gene linked to the CaMV35S promoter resulted in transgenic plants in which resistance to the fungus correlated directly with chitinase activity. It will be interesting to determine whether chitinase gene expression confers cross-protection to other fungal pathogens. More recently, the introduction of the stilbene synthase gene, which is thought to be involved in the synthesis of a phytoalexin, provided protection in rice to infection by the fungus *Pyricularia oryzae*. One of the challenges facing biotechnologists is to modify plants so as to increase net carbon gain. $C_4$ plants, such as maize and several weed species, have evolved a biochemical mechanism to overcome oxygen inhibition of photosynthesis.

In an initial assessment of the feasibility of improving photosynthesis in $C_3$ plants, the intact gene of *phosphoenol-pyruvate* carboxylase (PEPC), which catalyses the initial fixation of atmospheric carbon dioxide in maize, was introduced into japonica cultivars of rice, a $C_3$ plant. Transgenic plants exhibited reduced oxygen inhibition of photosynthesis and photosynthetic rates comparable to those of non-transformed plants. Such an approach for modifying one of the major physical processes plants holds promise for the transformation of other $C_3$ crops. Other experiments have been reported which attempt to increase the resistance of crop plants to environmental stresses such as ozone, high light, drought, cold and heat. A common feature in stressed plants in the production of free oxygen radicals which damage DNA, lipids and proteins.

In this respect, transformation procedures have been presented to increase the levels of superoxide dismutase, ascorbate peroxidase and catalase in cells in order to improve the tolerance of maize and rice to oxidative stress. Experiments have been directed towards modifying the nutritional quality of rice grain. For example, introduction of a fatty acid desaturase gene from tobacco into rice resulted in modification of the proportions of linoleic acid and linolenic acid in fatty acids, with the former decreasing and the latter increasing, respectively. A significant recent advance has been the transformation of rice to produce beta-carotene, a precursor of vitamin A.

Thus, the introduction of genes for phytoene synthase, phytoene desaturase, carotene desaturase and lycopene cyclase from *Narcissus*, or a double-desaturase from the fungus *Erwinia uredovora*, resulted in transgenic plants with grain producing yellow endosperm. Some lines produced enough beta-carotene to supply the daily human requirement from 300 grams of cooked rice. Since vitamin A deficiency affects about 7% of the world population (mostly children), mainly in developing countries, this work represents a significant advance in attempts to alleviate the problem of vitamin A deficiency. In the future, it may be essential to engineer complex biochemical pathways by the introduction of several transgenes into target species. In order to provide a foundation for this technology, embryogenic tissues of rice have been bombarded with a mixture of 14 different genes on pUC-based plasmids. Eighty per cent of the regenerated plants contained more than two and 17% more than nine of the transgenes. Importantly, plants with transgenes were phenotypically normal and 63% set viable seed. Detailed information collected over several seed generations from such plants will clarify the interaction and expression of multiple transgenes in genetically engineered plants, such as cereals.

## Summary: Problems, Limitations and Future Trends

Despite progressing from basic research involving the development of transformation protocols to the introduction and expression of useful genes in several rice and maize cultivars, cereals transformation technology still has its limitations. For example, technology transfer between laboratories and between individual workers is particularly difficult in cereals. Additionally, reduced expression or failure of transgenes to express, is reflected in low transformation frequencies. Failure of transgenes to express may be related to positional effects associated with the integration of foreign DNA into the recipient plant genome and to transgene silencing. Transgenes silencing is caused by cytosine methylation, an inherent control mechanism for gene expression. It is often associated with sense-suppression of multiple integrated copies of the transgene, or to co-suppression of an endogenous gene.

In rice transformed by particle bombardment, Kohli et al. presented evidence that truncation of trangenes produces incomplete transcripts giving aberrant RNA species, which may also be responsible for transgene silencing. Other workers have also shown that transgene silencing in rice occurs at the transcriptional level, although no evidence was found for gross alterations or methylation of CCGG sites in a chitinase transgene driven by the CaMV35S promoter. Kumpatla and Hall have stressed the need for detailed molecular analyses to be performed over several seed generations on plants harbouring transgenes and that lines containing methylated inserts should be carefully evaluated before being included in long-term breeding strategies.

Plants transformed at a single locus should be more amenable to breeding programmes, as the single transgenic locus will be easier to characterize genetically. These authors stated that particle bombardment generally results in a single transgenic locus, with this site acting as a 'hot spot' for subsequent integration of successive transforming DNA molecules. Currently, there is no simple procedure for the rapid identification of transgenic plants. Indeed, detailed molecular analysis, usually involving Southern blotting procedures of the DNA isolated from individual plants from large populations of regenerants, is still essential in order to identify individuals with single copy gene inserts.

Transformation with vectors carrying scaffold attachment regions (SARs) or matrix attachment regions (MARs), chromatin boundary elements that insulate genes from the effects of the surrounding chromatin, should assist in reducing the variability of transgene expression in transgenic plants. Other problems associated with the transformation of cereals include reduced access to new technologies, genes and promoters due to intellectual property and patents rights, which limit research progress. It is clear that attention must e paid to the gene delivery technology to minimize genomic and/or phenotypic effects. In this respect, Arencibia et al. performed detailed molecular analyses on rice transformed by particle bombardment and intact cell electroporation and concluded, importantly, that their transgenic plants exhibited negligible genomic changes as a result of the transformation process. They emphasized the need to select appropriate techniques to generate plants which express foreign genes of interest, whilst retaining existing agronomic and/or industrial traits.

Some of the recent commercial release of transgenic plants are listed by Birch and discussed by Dunwell. Before such release are possible, field analysis is compulsory to determine genetic stability, long-term effectiveness in the field and to assess any potential hazards associated with transgenic plants and their products. Presently, there is considerable public concern for the risks of genetic engineering, especially in Europe and the UK. Some of the potential problems associated with the technology have been summarized by Boulter and Rogers and Parkes. Such risks include the establishment of weedy, feral crop populations, transgene escape to wild species through sexual hybridization to produce new and uncontrollable herbicide-resistant weeds, pests or diseases, and the production of novel allergenic or toxic compounds by transgenic plants.

The escape of transgenes by pollen dispersal and through sexual hybridization can be prevented by plastid transformation, since these organelles are not transmitted, in most plants, through pollen. However, the success of this procedure will necessitate the development of systems in which gene constructs are targeted to plastid DNA. Thee authors have also provided evidence that certain genes, such as *Bt* genes for insect resistance, are expressed to higher levels in plastids than following nuclear integration. It should not be overlooked that many of the potential risks relating to genetic engineering of crops, which are currently being discussed, can also be associated with traditionally bred cultivars. It is of interest that the latter have been in existence for long periods without any cause for concern.

Risks are also posed by the cultivation of a homogeneous crop containing a useful gene, such as the *Bt* gene. This will increase selection pressure generating *Bt* toxin-resistant insect biotypes, leading to the rapid development of insects which can tolerate the *Bt* toxin and which are, as a consequence, able to overcome the resistance of transgenic plants. Strategies to reduce this risk include the use of several version of the *Bt* toxin gene, reducing selection pressure for the development of resistance to a single form of the toxin. Companies generating herbicide-resistant crops emphasize that the use of transgenic plants will reduce normally damaging and polluting herbicide applications by encouraging the use of safer, biodegradable herbicides. However, the availability of plants with resistance to a specific herbicide could lead, conversely, to overuse of the herbicide. A reduction in the broad spectrum of herbicides currently employed in agricultural practice could stimulate the development of herbicide-resistant weeds, as a result of selection pressure.

The use of antibiotic-resistance genes to select transformed cells prior to the regeneration of transgenic plants, also poses potential hazards of horizontal gene transfer, including the risk of transfer of the antibiotic resistance genes to bacteria in the gastrointestinal tract. The production of marker-free transgenic plants is becoming an important consideration as a consequence of the concerns associated with herbicide and antibiotic resistance genes. Yoder and Goldsbrough detail the approaches available in marker-free technology. In summary, methods exist to excise a marker gene from transgenic plants, including the use of prokaryotic site-specific recombination systems, such as the bacteriophage P1 *Cre/lox* system, where a transgenic plant is generated containing two lox sequences flanking the selectable marker. The *Cre* gene, introduced through a second transformation, catalyses recombination between two lox sequences, resulting in excision of the undesirable marker gene.

Transposable elements can also be used to excise marker genes in a similar manner. For example, the maize *Ac* transposable element encodes for transposase, which trans-activates the *Ds* transposable element causing its excision and reinsertion into a new locus. Any marker gene sequences cloned between the inverted repeats of a *Ds* element, are also activated. Studies have shown that 10% of excised elements do not reinsert, leading to loss of the marker gene. Perhaps the most simple approach for generating marker-free transgenic plants is that of co-transformation. Thus transformation with two plasmids, with the undesirable marker gene on a separate plasmid from the useful gene, will allow selection of a marker-free plants following segregation in seed progeny. The success of this process requires that the two transgenes are co-transformed efficiently and into unlinked sites.

Finally, tissue- or temporal-specific expression will permit control of marker gene expression in transgenic plants. Current objectives for cereal transformation must be response to the requirement of plant breeders who are directed by the product consumer. Such objectives include improvements in crop yield, pest and disease resistance, stress tolerance and nutritional quality. These improvements will necessitate the identification and cloning of agronomically useful genes prior to their introduction into specific crops. Undoubtedly, one of the major research objectives must remain the manipulation of cereals to fix their own nitrogen since this would have considerable environmental impact by reducing the application of nitrogenous fertilizers normally required to maximize crop yield.

Improvements in vector design and construction are still required in order to generate transgenic plants in which single copies of the gene(s) of interest are inserted precisely into the plant genome and which express reliably throughout seed generations. Overall, transformation procedures should be simple, in order to facilitate technology transfer, and independent of plant genotype. Such transformation technology must be underpinned by reliable tissue culture procedures in order to identify those cells which are most competent for transformation and which are capable of regenerating into phenotypically normal, fertile transgenic plants. In addition, the time of culture should be minimal. The nature and timing of the culture parameters are probably more important for the recovery of transgenic cells and plants than the selection procedure. At present, particularly in the UK, consumer acceptance of genetically engineered crops, including cereals, poses one of the chief limitations to the exploitation of the technology discussed in this chapter. Clearly, the introduction of transgenic crops for human consumption must proceed slowly and with caution in order to accumulate reliable data on questions, such as the frequency of transgene introgression from transgenic crops to wild species, in order to build up public acceptance and confidence in the safety of such products. This should be combined with improved communication, by workers actively involved in gene manipulation technologies, of a balanced and unbiased view regarding transformation technology, in order to bridge any gaps in knowledge between scientists and the public.

## Sources of Further Information and Advice

The reference cited in the text are the prime source of information relating to the transformation of rice, maize and other cereals. The review articles are particularly helpful in this respect, since they provide detailed overviews of the subject. All published articles carry the address of the workers who have been responsible for conducting the experiments and collating the data. Most authors will provide detailed information, upon request, relating to aspects of their published procedures; many welcome visits to their laboratories to facilitate technology transfer at the practical level. In addition, it is now the policy of most international journals to include the e-mail address, fax and the telephone numbers of communicating scientists in order to facilitate communication in this rapidly advancing field of research. Several internet sites are also available, which are useful in keeping workers informed of the current advances in the field of transgenic research in relation to crop improvement. For example, the Bowditch Group produces a News Bulletin to update researchers on the latest developments, particularly by agrochemical companies, in plant biotechnology, while the University of Amsterdam also collates similar information.

# Chapter 11

# Transgenic Cucurbits and other Vegetables

This chapter is related to the biotechnology of a number of vegetables that are botanically and morphologically unrelated but that have been associated because they are generally secondary crops on the world scale as compared to the species dealt with in the previous chapters. However, some of them may have great importance in some areas as they provide an essential part of the diet and/ or an important income complement for farmers.

The group includes:

1. Bulky organs other than potato corresponding to roots (carrots), tubers (sweet potatoes) and bulbs (Allium species such as onion and garlic).
2. Leafy vegetables including lettuce, spinach and brassica species (cabbage, broccoli, cauliflower, etc.).
3. Fleshy fruits of the cucurbitaceae (melon, squash, cucumber) and solaneaceae (pepper, eggplant) families that are eaten raw or after cooking.
4. Legumes other than soybean that are eaten as vegetables such as peas and beans.

Overall these crops have received less attention in terms of genetic engineering than other major crops such as maize, rice or soybean or than model plants such as *Arabidopsis* and tobacco, although progress in the transformation procedures in major crops or model plants has been beneficial to secondary species. Nevertheless, some of the species covered in this review remain recalcitrant to transformation and in many cases, although successful gene transfer has been claimed, the efficiency is too low for routine transformation and/or is restricted to certain cultivars.

## Biotechnology of Cucurbits

Cucurbits is a general term for the species of the Cucurbitaceae family. This family comprises about 130 genera and more than 900 species of which only a few are cultivated. Melon, watermelon, pumpkins, squash, gourds and cucumber represent the most important cucurbits. The world production of these crops is in excess of 115 million metric tons, on a total area harvested of 6.9 million hectares throughout the world. The world production and the harvested area has increased three-fold in the last three decades. Similarly, yields have been also greatly improved in the last few years, for instance from 11.6 tons/ha in 1970 for cucumber to 16.2 tons/ha today. In this chapter, we concentrate on the most widely cultivated species with emphasis on those for which biotechnology methods have been applied.

### Methods of Transformation

Efficient transformation methods require good control of the regeneration step through either organogenesis or somatic embryogenesis. In cucurbits, regeneration by direct or indirect organogenesis has been achieved from various plant tissues including cotyledons, hypocotyls and leaves. The efficiency of regeneration has been found to be highly dependent on the stage of development and the growth conditions and the genotype. Regeneration of cucurbits has been obtained by somatic embryogenesis from callus or different explant tissues, but development of somatic embryos into plants still remains difficult, being highly dependent on the genotype and controlled by crossing. This renders the process of plant transformation more difficult and reduces the number of cultivars susceptible to be engineered.

On the other hand, the regeneration process is known to induce endo-polyploidisation leading to reduced productivity of the plants and altered shape and smaller size of the fruit, mainly in melon. Endo-polyploidization occurs during the development of cotyledons, and at a lower rate in leaves. Thus, the regeneration protocols must be adapted to yield low endo-polyploid plantlets. Currently, several methods for gene transformation including *Agrobacterium*-mediated transformation and biolistic are available.

A great number of melon, cucumber and squash genotypes have been transformed with different genes and in particular with genes of agronomic interest.

Transgenic plants currently tested in field conditions or commercialized were generated via *Agrobacterium*-mediated transformation. In cucurbits, cotyledonary explants and adventitious shoots have been widely used for generation. In these transformation systems, the *NptII* selectable marker gene is the most widely used allowing selection of transformed tissues in the presence of kanamycin. Biolistic transformation has been used for the generation of transgenic cucumber and melon plants, respectively from highly embryogenic cell suspension cultures or embryos developed on cotyledons leading to a high stability of the transgene. Genetic transfer via *A. rhizogenes* in cucumber or *Cucurbita pepo* or via the pollen-tube pathway in watermelon have been described but so far do not seem to have provided an efficient process in cucurbits.

**Resistance to Viruses**

More than 30 viruses can infect cucurbits, but only twelve cause economically important losses. The following viruses are the most commonly encountered in warm and temperature areas and capable of infecting several cucurbit species:

(i) CMV (Cucumber Mosaic Virus);
(ii) WMV (Watermelon Mosaic Virus);
(iii) ZYMV (Zucchini Yellow Mosaic Virus),
(iv) PRSV (Papaya RingSpot Virus); and
(v) SqMV (Squash Mosaic Virus).

The incorporation of virus resistance genes into cucurbits has been the goal of many breeding programmes. However, because resistance genes are derived from wild species, they are not simply inherited and/or are recessive, so introgression of these genes into horticulturally acceptable genotypes is not an easy task. Genetic engineering has allowed breeders rapidly to develop virus-resistant varieties by introducing dominant coat protein or replicase viral genes into inbred parents of existing commercial hybrids. Viral coat protein-mediated protection has allowed the generation of CMV-resistant cucumber plants that can exhibit a high level of CMV-resistance under field conditions. Similarly, melon plants over-expressing CMV or ZYMV coat proteins have been generated.

For Clough and Hamm, a significant reduction in disease incidence in the transgenic lines occurred in field conditions. Expression of the coat protein gene delayed virus disease development and the subsequent systemic spread of virus in the transgenic plants. However, because most cucurbits are susceptible to several viruses, multiple virus resistance has been sought, instead of single virus resistance.

Transgenic lines of yellow crookneck squash (Cucurbita pepo) containing multiple coat protein constructs of CMV, WMV and ZYMV have been generated and used for the production of hybrid varieties. Field evaluation was performed for transgenic lines:

(i) CZW-30, transformed with the triple coat protein gene construct that exhibited resistance to all three viruses and

(ii) ZW-20, transformed with the coat protein genes of WMV and ZYMV that displayed excellent resistance to the two viruses.

The ZW-20 line and subsequent generations were approved for commercial distribution by Asgrow in the USA in 1995 and were the first disease resistance transgenic plants to be approved for commercialization. Similarly, two experimental transgenic summer squash hybrids, possessing resistance to ZYMV and WMV and to ZYMV, WMV and CMV exhibited outstanding resistance in field conditions to the corresponding viruses as compared to the non-transgenic virus-susceptible hybrid 'Pavo'. Another strategy to enhance virus tolerance in melon, based on the over expression of polyribosome directed toward CMV coat protein, has been reported by Plages. A further approach to reduce development of the virus in cucurbits is to use a cDNA copy of RNA1 or an altered form of the 2a *replicase* gene from CMV which have been found to enhance virus tolerance in transgenic tobacco.

**Resistance to Fungi**

The improvement of resistance to phytopathogenic fungi is one of the most crucial objectives in cucurbit cultivation. However, the problem is complex because highly resistance sources must be available for the breeding programmes. For example, *Fusarium* wilt-resistant materials have been found in genetic resources in cucumber and used in breeding programmes. On the contrary, breeding materials for resistance against grey mould (*Botrytis cinerea*), one of the most serious cucumber disease have not been found. Transformation techniques have been used to produce resistant transgenic plants by expressing chitinase genes aimed at inhibiting fungal development in the plant. In cucumber, the response of transgenic plants expressing different chitinase genes originating from petunia, tobacco or bean to inoculation with fungal pathogens including *Alternaria radicini*, *Botrytis cinerea*, *Colletotrichum lagenarium*, and *Rhizoctonia solani* has shown that chitinase overexpression failed to reduce the development of the disease. In opposition, Tabei et al. have succeeded in obtaining *B. cinerea*-resistant lines of cucumber by expressing a rice endochitinase gene.

Different responses for disease resistance were observed including inhibition of appressorium formation and penetration of hyphae as well as restriction of invasion of the infection hyphae. Furthermore, disease resistance against grey mould was confirmed to be inheritable and chitinase overexpression did not affect the morphological development of the plants. Although the effectiveness of the fungal resistance must be confirmed by field trials, this approach could be used for other cucurbits species to fight against plant diseases caused by phytopathogenic fungi. However, the extent of disease tolerance appears to be highly dependent on the type of chitinase protein expressed and the characteristics of the fungal pathogen. On the other hand, chitinase overexpression in combination with a-glucanase, PR proteins or other enzymes involved in the synthesis of antifungal compounds could provide a broader spectrum of activity against fungal pathogens.

**Resistance to Abiotic Stresses**

Unfavourable soil and water conditions are important constraints and lower crop yields. The adaptive response of plants to water and salt stress involves the activation of a large number of genes. The individual function of these genes is not well understood in as much as most of the genes are not specific to these types of stresses. The number of plants that have been engineered for salt or drought resistance is therefore very limited. However, a yeast salt-tolerance gene encoding a water-soluble protein HAL1 has been transferred to two cultivars of *Cucumis melo* via *A. tumefaciens*. Some halotolerance has been observed in primary transformants that could be transmitted to self-pollinated progeny.

**Fruit Quality Traits**

In recent years, methods based on the genetic manipulation of fruit ripening, have been developed as an alternative strategy to increase the storage life and improve the quality of fruits. Although most studies have been carried out with the tomato as model fruit, Cantaloupe melons represent good targets due to their fast ripening rate and short post-harvest life.

Post-harvest losses largely due to over-ripening have been estimated to be near 30% in the USA. In climacteric fruit like melon, the expression of many genes involved in the ripening process is stimulated by ethylene. This plant hormone represents an obvious target for controlling fruit ripening by genetic manipulation. An antisense construct of a cDNA encoding melon ACC oxidase (ACO) driven by the 35S promoter has been used to generate transgenic melons of the Cantaloupe Charentais type (cv Vedrantais). Among several transformants, one line was selected that exhibited strong inhibition (over 99.5%) of ethylene production. Ethylene suppression respiration and peduncle detachment. However, coloration of the flesh, accumulation of sugars and organic acids, and the synthesis of the ethylene precursor ACC were not affected by ethylene suppression..

Ethylene-inhibited fruit also evolve far fewer aroma volatiles than wild-type (WT) fruit. However, ethylene treatment of antisense ACO fruit is capable of restoring the WT phenotype. Antisense ACO charentais cantaloupe melons enabled an assessment of the role of ethylene in some physiological disorders. These melons, in contrast to WT fruit, do not develop the characteristic pitting and browning of the rind associated with chilling injury either when stored at low temperature (2°C for 3 weeks) or upon rewarming to room temperature. Tolerance to chilling was clearly correlated with a lower accumulation of ethanol and acetaldehyde and a higher activity of activated oxygen scavenging enzymes in antisense ACO fruit. It appears that ethylene acts in conjunction with low temperature to induce metabolic shifts that participate in the development of chilling injury. Another strategy for reducing ethylene synthesis has been used for cantaloupes of the American type commonly referred to as muskmelon.

The T3 bacteriophage gene product S-adenosylmethionine hydrolase (SAMase) catalyses the degradation of SAM, a precursor to ethylene biosynthesis. The gene was expressed under the control of a chimerical fruit-specific promoter. It was confirmed that the chimerical promoter was capable of driving SAMase expression in a fruit-specific and ethylene-responsive manner. SAMase melons showed significant reduction in ethylene biosynthesis (up to 75%)

both as inbred homozygous plants and as hybrids. The inhibition of ethylene production of SAMase fruit was not strong enough to dramatically alter the ripening and Post-harvest phenotype.

However, although the onset of maturity was not significantly delayed, full maturity of transgenic fruit occurred over a shorter period. Also, the concentration of soluble sugars was frequently higher in transgenic fruit probably because fruit slip was delayed by one to three days, allowing more sugars to accumulate in the fruit before its harvest. These examples indicate that two alternative exist for the Post-harvest handling of ethylene-inhibited Cantaloupe melons: (I) use of lines and corresponding transgenic hybrids with very strong inhibition of ethylene production (in this case fruit can be stored in the cold with no risk of chilling injury and then ripened to the desired stage of maturity before distribution), or (ii) use of lines and corresponding hybrids with less reduction of ethylene production and showing delayed ripening and extended shelf-life so as to reach the consumer at the right maturity stage without ethylene treatment. A comparative evaluation of the ethylene-inhibited transgenic genotypes and already existing long-keeping varieties of melon deserves to be carried out form the point of view of quality of the fruit, storability and Post-harvest shelf-life.

## BIOTECHNOLOGY OF PEPPER

Peppers belong to the genus *Capsicum*. They are found around the world in various edible forms and they exhibit great variations in size, shape, flavour and colour as well as plant habits. Five main species have been domesticated (*C. frutescens* L., *C. chinense* Jacq., *C. baccatum* L., *C. pubescens* R.& P., and *C. annuum* L.) among which *C. annuum* is the most widely cultivated. Two types of cultivars can be distinguished: sweet and pungent. A horticultural classification of peppers has been made by Smith based on the size, shape colour and taste (sweet or pungent) of the fruit. The transfer of characters from wild species to cultivated genotypes is hampered by interspecific incompatibility and/or hybrid sterility. Only in a few cases have these barriers been overcome through embryo culture and somatic hybridization. A recent update on pepper *in vitro* regeneration and transformation has been made by Steinitz et al.

### Methods of Transformation

Most of the reports on pepper regeneration deal with two main types, bell and chile. Shoot regeneration is generally dependented on organogenesis from cotyledons and hypocotyls, but the regeneration capacity is highly dependent upon the cultivar, the developmental stage and the location of the plant tissue. Zhu *et al.* found that leaves of a Chinese sweet pepper variety were by far the best material for successful regeneration, transformation with young leaves giving the highest rates. However, Arroyo and Revilla had previously described an efficient regeneration procedure for some commercially important Spanish cultivars using hypocotyls and especially cotyledons with high regeneration rates. Fari *et al.* by screening a seed collection of chile peppers for *in vitro* regeneration selected plants that, after selfing, generated an inbred line n°40017-13 with a high capacity for regeneration.

Regeneration from protoplasts has also been achieved in one cultivar of *C. annuum* 'Dulce Italiano', while three other cultivars and a wild species *C. chinense* produced meristem-like structures but no shoots. Regeneration of chile peppers has also been achieved by somatic embryogenesis. Concerning bell peppers, as *in vitro* regeneration protocol including transformation via *A. tumefaciens* harbouring a GUS reporter gene has been defined by Liu *et al.* using six cultivars and one wild accession. Regeneration of whole transgenic plants proved unsuccessful. A critical step appears to be the elongation and rooting of the shoots during kanamycin selection.

The virulence of *Agrobacterium* strains towards peppers seems to be variable and dependent on the cultivar, using leaves as starting material and a specific protocol, specially for bud elongation, have succeeded in generating a Chinese sweet pepper variety harbouring a CMV coat protein. Concerning chile peppers, Manoharan *et al.* established a protocol for regeneration-transformation of a hot chile pepper variety from India using a hypervirulent *A. tumefaciens* strain (EHA 105) and involving the use of thidiazuron as a cytokinin as suggested by Szasz et al. for bell peppers.

An effective protocol has also been established for a Korean variety of hot pepper that uses a complex cocktail of plant growth regulators and a pre-culture of cotyledons or hypocotyls in the presence of the ethylene inhibitor $AgNO_3$. Despite the elevated number of protocols published so far, their efficiency is still very low and peppers can still be considered as recalcitrant to genetic transformation. It is obvious that a good combination of an efficient protocol for regeneration-transformation, a virulent *A. tumefaciens* strains, and highly responsive cultivars is the key for the generation of transgenic peppers.

**Transformation for Herbicide Resistance**

Transgenic hot peppers showing resistance to bialaphos, a non-selective herbicide have been generated by transfer of the *bar* gene via *A. tumefaciens*. Transgenic plants were capable of withstanding 5000 mg.$L^{-1}$ of bialaphos applied to the leaves. Sweet peppers expressing the pat (phosphinothricin N-acetyl transferase) gene introduced via *A. tumefaciens* exhibited tolerance to application of 0.44% of the commercial preparation of Basta containing 20% phosphinothricin.

**Resistance to Virus**

Transgenic sweet peppers harbouring a CMV satellite RNA showed increased resistance to viruses. Stable inheritance of the gene was observed in the progeny and significant symptom attenuation of the offspring was confirmed upon mechanical inoculation with CMV-Y or CMV-Kor strains under greenhouse conditions. However, feasibility of satellite RNA as a biocontrol of CMV in hot peppers needs further field study experiments.

**Peppers as a Source of Gene and/or Promoters of Interest for other Crops**

Peppers have been used as a model for the study of carotenoid biosynthesis. Two genes that encode major chromoplast proteins have been cloned that encode a capsanthin/capsoburin synthase (*ccs*) involved in the synthesis of a red carotenoid pigment not found in tomato and a fibrillin (*fib*), structural protein involved in carotenoid deposition in chromoplasts. Both genes (*ccs* and *fib*) are strongly induced at the early stage of ripening in peppers. The two promoters of the genes, specially the *fib* promoter were exceptionally strong in expressing GUS activity in ripening tomato fruit starting at the late immature-green stage. Interestingly, although ethylene is considered not to be involved in the ripening of the non-climacteric pepper fruit, it influenced GUS expression driven by both the *ccs* and *fib* promoters in the climacteric tomato fruit.

## Biotechnology of Eggplant

Eggplant and its cultivated or wild relatives, cover a wide range of *Solanum* species (subgenus *Leptostemonum*) originating from

Asia and Africa. The most widely cultivated species in *S. melongena* L. Other species are cultivated in Africa such as the scarlet eggplant (*S. aethiopicum* L.) and the gboma eggplant (*S. macrocarpon* L.). Eggplant (*S. melongena* L.), with a world production of around 9 million metric tons, is an economically very important solanaceous crop in Asia and the Mediterranean basin.

**Methods of Transformation**

Early work by Guri and Sink, Rotino and Gleddie and Rotino et al. demonstrated the feasibility of transformation of eggplant via *A. tumefaciens* with a transformation efficiency of 7% in the latter case. However, Billings *et al.* were unsuccessful in transforming their experimental material using one of these two methods. After studying the effects of growth regulators, they devised an improved method that included thidiazuron and gave an transformation efficiency of 20.8%. The transformation efficiency seems to depend strongly upon the genotype. For instance, La Porta *et al.* were unable to transform their lines using the protocol of Billings *et al.*, which proved very efficient with some other genotypes, while they obtained a transformation frequency of 8.3% with the protocol of Rotino and Gleddie. It appears therefore that eggplant is not a recalcitrant species, but care must be taken in adapting the transformation protocol to the genotype.

**Transformation for Herbicide Resistance**

No genetic transformation aimed at conferring herbicide resistance to eggplant has been published so far. However, eggplants resistant to atrazine have been generated by screening somatic embryos derived from EMS-mutagenized seeds for their ability to grow in the presence of 15 mg.L$^{-1}$ of atrazine.

**Resistance to Insects**

The commercial production of eggplants is frequently hampered in western countries by attacks by Colorado potato beetle (CPB) (*Leptinotarsa decemlineata* Say). Because the eggplant germplasm lacks an effective resistance gene to CPB, improvement of insect resistance via biotechnology represents the only alternative for the generation of insect-resistant eggplants. The insecticidal crystal protein genes (*Cry*) of *Bacillus thuringiensis* Berl. represent an important gene family for the biotechnological improvement of resistance to certain insects (mainly lepidoptera and coleoptera) in cultivated plants. The native *CryIIIB* gene was first used to produce eggplants resistant to the CPB following the transformation protocol described by Billings et al. A study of the seedlings of eight independent transgenic lines revealed the absence of any significant resistance to the first and second instar larvae of the CPB. Using a modified synthetic *CryIIIA* gene several lines of transgenic eggplants were generated that now showed resistance to neonate larvae and adult CPB under filed conditions. These lines showed higher expression of the *CryIIIA* gene than the previously tested native *CryIIIB* gene.

A mutagenized version of the *CryIIIB* gene has been transferred by Arpaia et al. to the female parent of the commercial eggplant F1 hybrid 'Rimina'. Over 150 transgenic plants were produced among which 23 showed high expression of the toxin and significant insecticidal activity on neonate CPB larvae. Further tests were done on selfed transgenic progeny showing significant resistance to CPB and higher yields under natural infestation in field conditions. In Asian countries the most devastating insect of eggplants is a lepidopteran fruit and stem borer (*Leucinodes orbonalis* Guenee). Kumar et al., using the transformation method of Rotino and Gleddie, have generated transgenic Brinjal type eggplants expressing a synthetic *CryIAb* gene. Plants strongly expressing the toxin showed significant insecticidal activity against the larvae in bioassay studies.

Glasshouse and field evaluations remain to be performed to critically evaluate the practical usefulness of the strategy. Altogether, these results demonstrate that recovery of a high level of CPB resistance can be achieved by using modified *Cry Bt* genes rather than native ones and that biotechnology is an efficient strategy for the control of CPB in eggplants. Another type of gene for plant resistance to insects has been tested. It encodes proteinase inhibitors that prevent digestion of plant proteins by some coleopteran and hemipteran insects. La Porta et al. have transformed eggplant using an optimized protocol of Rotino and Gleddie with a soybean gene encoding a cysteine inhibitor of proteases, but information on the resistance to insects of the transgenic plants generated is lacking.

## Quality Traits

The absence of seeds is a desirable trait in a number of fruit crops including eggplant. Parthenocarpic mutants have been identified in several plant species, but their use for generating parthenocarpic varieties is limited by the reduction of fruit set and fruit size. The parthenocarpic trait is polygenic and it proves cumbersome in breeding programmes. For these reasons biotechnology may prove to be an interesting alternative method. Transgenic eggplants expressing the coding region of the *iaaM* gene from *Pseudomonas syringae* driven by an ovule-specific promoter show parthenocarpic development. The *iaaM* gene codes for an indolacetamide monoxygenase that converts tryptophan to indolacetamide, a precursor of the plant hormone auxin. Transgenic plants produced seedless fruit of marketable size when the flowers were emasculated or in adverse conditions when untransformed lines where unable to set fruit.

The comparison of three hybrids, transgenic for the *iaaM* gene, with untransformed hybrids and a commercial parthenocarpic cultivar in winter in an unheated greenhouse showed that the yield of the transgenic hybrids was increased by ca. 25% even in the absence of fruit set hormone treatment. A 10% reduction in the cultivation costs was also observed. It is concluded that the *iaaM* gene is a powerful biotechnological tool for generating parthenocarpic eggplants that proves to be superior to the use of both agricultural practices and traditional genetic methods.

## BIOTECHNOLOGY OF LEGUMES

Legumes or pulses comprise a large number of species among which only a few are extensively cultivated. Christou has reviewed advances in grain legume biotechnology and reports on all legume species. Here we will concentrate on legumes that are mainly consumed as vegetables. The two major species are common beans (*Phaseolus vulgaris* L.) and peas (*Pisum sativum* L.). Reference will also be made to other species of lesser economical importance for which biotechnological improvement has been undertaken, such as cowpea (*Vigna unguiculata* L. Walp.) mungbean (*Vigna vradiata* L.), chickpea (*Cicer arietinum* L.), faba bean (*Vicia faba* L.) and lentils (*Lens culinaris* Medik).

### Methods of Transformation

#### *Peas*

Successful transformation of peas has been achieved through *A. tumefaciens*-based transformation vectors. Gene transfer through *A. rhizogenes* has also been reported but with the aim of studying expression of transgenes during hairy root development. Direct DNA transfer into protoplasts using electroporation has resulted in the generation of transgenic calli that were resistant to hygromycin, but plant regeneration from these cultures was not possible.

Production of transgenic peas via *A. tumefaciens*-mediated gene transfer has been achieved by Puonti-Kaerlas et al. Studies into the inheritance of the transgene gene showed stable transmission of the gene over two generations, but some aberrations were reported in the primary transformants including aborted flowers, limited number of seeds per pod, non-viable seeds and polyploidy. The nine-month period required to recover transgenic shoots is a sing of low efficiency of the protocol and is probably responsible for the aberrations observed. More recent protocols requiring shorter times have allowed further advances to be made towards routine transformation of peas.

Schroeder et al. have described a protocol that used longitudinal slices of immature embryos and phosphinotrocin as a selectable agent in which 1.5% to 2.5% of the starting explants gave rise to normal transformed plants. Grant et al. have developed a method using immature cotyledons as the explant source in which both organogenesis and embryogenesis regeneration seems to occur. The method was applied to four distinct cultivars. It allowed reduction of the time from explant to seed-bearing primary regenerants. Further assessment of this method demonstrated the efficiency of kanamycin in selecting transformed peas. Davies et al. obtained transgenic plants by injecting *A. tumefaciens* into the cotyledonary node. The selection agent was kanamycin and approximately 1.4% of the injected seeds gave rise to transgenic plants. The method was further modified for greater reliability. This procedure presents distinct advantages over those previously reported in that it uses dry mature seeds as starting material and cotyledonary meristems producing shoots without an intermediate callus phase. It allowed rapid generation of phenotypically normal self-fertile plants in which transgenes were inherited in a Mendelian fashion.

### Beans

Although considerable efforts have been made to establish regeneration protocols by direct shoot formation from apical and axillary meristems or from embryo-derived calli, an efficient

*Agrobacterium*-mediated transformation system has not yet been developed. Reports claiming transformation of *Phaseolus vulgaris* are not convincing as they lack genetic and molecular analysis of the transformants to confirm stable transformation. The slow progress in *P. vulgaris* transformation using *A. tumefaciens* could be attributed to the lack of knowledge about the sensitivity of bean genotypes to *A. tumefaciens* or factors affecting transformation itself.

Direct gene transfer by particle bombardment has proved more successful. Using a protocol similar to that developed for soybean with embryo axes generated plants that expressed GUS activity and resistance to the phosphinothricin herbicide over several generations. Kim and Minamikawa also transformed beans using particle bombardment. They were able to introduce the *GUS* gene driven by the concanavalin A promoter conferring seed specificity. The protocol allowed the generation of six transgenic plants from 319 embryogenic axes. Stable integration was confirmed only in primary transformants and not in the progeny. The most obvious success of the particle bombardment process in generating transgenic beans has been reported by Aragao et al. who generated methionine-enriched and virus resistant beans in which the transgene was proved to be stably transmitted for at least three generations.

### *Broad bean (Vicia)*

An *A. tumefaciens*-mediated transformation protocol including the generation of embryos from shoot tip calli has been devised for *Vicia narbonensis*, a close relative of the faba bean (*Vicia faba*). It has allowed the generation of stable transgenic lines producing methionine-enriched seeds.

### *Mung beans (Vigna)*

The particles delivery system has been used to express the *GUS* and *NptII* genes in three in three *Vigna* species (*V. radiata*, *V. aconitifolia*, and *V. mungo*) at a relatively high frequency. Transformed calli of *Vigna mungo* and roots of *Vigna radiata* have been obtained using *A. tumefaciens* and non-disarmed *A. rhizogenes* respectively, but none was capable of regenerating shoots. Similarly, attempts to transform cowpea with *Agrobacterium* failed to generate transgenic plants. A more successful method has been reported by Chowrira et al. who obtained transgenic cowpea by electroporation-mediated gene transfer into intact plant tissues: the *GUS* gene was introduced in electroporated nodal axillary buds and R1 plants were recovered from seeds originating from these buds.

### *Lentils*

Very few attempts have been made at the genetic transformation of lentils. Early work with four strains of no-disarmed *A. tumefaciens* showed that transfection was feasible, but the generation of transgenic plants was not reported. The possibility of transferring genes in cotyledonary nodes via particle bombardment has been explored recently by Oktem et. al. Stable expression of the GUS reporter gene was achieved in regenerated shoots. Using the same electroporation method as cowpea, Chowrira et al. obtained R2 plants expressing the *GUS* gene. However, segregation ratios in R2 population showed a strong bias against transgene presence or expression.

### *Chick pea*

Successful transformation and regeneration of the chickpea has been reported using embryo axes after excision of the apical meristem co-cultivated with *A. tumefaciens*. Transgenic plants expressing GUS and NPTII genes were obtained. Altinkut et.al. used shoot primordia of mature embryos and reported and Agrobacterium-mediated transformation with an efficiency of 12.7%. Particle bombardment on embryo axes has been used as alternative and allowed the production of plants expressing a chimeric *CryIA*(*c*) gene.

## Transformation for Herbicides Resistance

The *Bar* gene, conferring resistance to Bialophos herbicide and the related compounds phosphinothricin and ammonium glufosinate, has been used as a selection marker both for pea and bean transformation. The transgenic pea and bean plants generated showed strong resistance, in greenhouse conditions, to level of Basta similar to those used in field practice. Stability of the transgene was proved in the progeny.

## Resistance to Viruses

Bean golden mosaic geminivirus, transmitted by the white fly (*Bemisia tabaci*, Gen) is the causal agent of a severe disease of the common bean throughout western tropical regions. Only low moderate resistance can be found in the germplasm so highly resistant varieties cannot be generated by breeding. The potential of genetic engineering to produce geminivirus-resistant plants has been demonstrated in the tomato. Antisense constructs of the *Rep-TrAP-REn* and *BC1* viral genes have been used to transform common bean via the biolistic process using embryo apical meristems. Two of the transgenic lines from the R3 and R4 generations exhibited both delayed and attenuated viral symptoms, but not full resistance. Another approach has been made by Chowrira et al. with transgenic peas expressing a chimeric pea enation mosaic virus coat protein gene. R2, R3 and R4 plants were shown to display attenuated symptoms as compared to controls.

**Resistance to Insects**

The stored grains of peas and other legumes such as chickpea and cowpea are susceptible to storage insect pests, mainly bruchid beetles. The presence of an α-amylase inhibitor in bean seeds confers insect resistance and the *α-AI-PV* gene is therefore a candidate gene for the improvement of other legumes. Transgenic peas expressing the *α-AI-PV* gene under a seed-specific promoter (phytohaemagglutinin) were produced and seeds of the R2 generation caused increased mortality in Azuki bean weevil (*Collossobruchus chinensis*) and cowpea weevil (*C. maculatus*) larvae. Moreover, these transgenic peas showed strong resistance to the pea weevil (*Bruchus pisorum*) during seed development in the growing crop and were successfully tested under field conditions. Proteinase inhibitors (PI) are also involved in plant protection from insects and peas expressing a PI from *Nicotiana alata* have been obtained and were able to produce PI amounting to 0.2% of total protein. Feeding trials with R2 transgenic peas displayed showed development of *Helicoverpa armigera* larvae. Chick pea expressing a chimeric *CryAI*(*c*) gene from *B. thuringiensis* has been obtained by Kar et al. through particle bombardment. Insect feeding assays indicated an inhibitory effect on larvae of *Heliothis armigera*. Further studies on the inheritability of the gene and field behaviour of the transgenic plants are awaited.

**Quality Traits**

Although beans are rich in some essential amino acids (lysine, threonine, valine, isoleucine and leucine), their nutritional value is limited because of the small amounts of other essential amino acids methionine and cysteine. A *2S-albumin* gene has been isolated from the Brazil nut (*Bertholletia excelsa* H.B.K) encoding a methionine-rich protein (18.8% of total amino acids) which is targeted to the seed vacuoles. This gene has been expressed in a number of laboratory plants (tobacco, *Arabidopsis*) and seed crop species like canola, sunflower and potato. Expression in beans has been achieved using the biolistic method. The gene driven by a double 35S promoter and AMV enhancer sequences was stable and correctly expressed in homozygous R2 to R5 seeds. In two of the five transgenic lines, the methionine content was increased by 14 and 23%. The same gene has also been expressed in *Vicia narbonensis* using either the 35S or the seed specific promoter of the legumin B4 (*LeB4*) gene from *Vicia faba*. Transformation was performed via *A. tumefaciens* using this protocol of Pickard et al. The LeB4 promoter proved to be much more efficient than the CAMV 35 promoter. Besides conferring seed-specific expression, it induced a three-fold increase in the methionine content of seed protein in one R0 line. However, these studies are hampered by the fact that Brazil nut 2S albumin has been identified as an allergen and the transgenic soybeans expressing the protein were allergenic.

## Beans as a Source of Genes and/or Promoters of Interest for other Crops

Lectins are found predominantly in the seeds of legumes, where they accumulate to relatively high levels. Their role in defence against pests and pathogens has been established and lectin genes could be interesting candidates for agronomic improvement. Pea lectin was introduced into potato under CAMV promoter control. The lectin was processed in potato leaves as in pea cotyledons, and maintained its haemagglutination activity. The cowpea trypsin inhibitor has been used as an insectidal gene in tobacco and sweet potato. The pea storage protein legumin has been transferred to rice to improve the amino acid composition.

A number of pea storage proteins accumulate specifically in the seed. The corresponding genes exhibit seed-specific expression so that their regulatory sequences can be used to drive gene expression of heterologous proteins specifically in the seed. Such is the case for the promoters of *Phaseolus vulgaris* genes for arcelin, phaseolin, phytohaemagglutinin and a a-amylase inhibitor. The sequence motifs of the pea lectin promoter that confer seed-specific expression have been characterized. A *Vicia faba* legumin promoter has been used to express the 2S methionine-rich protein of Brazil nut in *Vicia narbonensis*. Legumes can also be a source of other tissue-specific promoters. The promoter of the *Blec4* gene from pea confers specific gene expression to the epidermal tissue of vegetative and floral shoot apices of alfalfa.

## Biotechnology of Bulky Organs (Carrots, Sweet Potatoes, Allium species)

Roots bulbs and other bulky organs used as vegetables belong to a wide range of horticultural species. Except for the potato, very few of them have received attention in terms of improvement via biotechnology. In this section we will review the present data available for some roots and bulbs. Most of them are related to two species.

1. The carrot (*Daucus carota* L.) which is cultivated world wide, representing 3% of world vegetable production (around 14 million tons annually).
2. The sweet potato (*Ipomea batatas* L.), which is one of the most important crops used for human consumption in developing countries and ranks sixth (around 115 million tons annually) among the most important crops in the world after wheat, rice, potato and barley. Sweet potatoes and carrots are both dicots.

Data on bulky monocotyledonous species are very scarce although Allium species (onion, garlic and leek) are important. This is mainly due to technical difficulties in transformation.

### Methods of Transformation

#### Carrot

Routine regeneration of carrot can be achieved through induction of somatic embryogenesis in cell or callus cultures. This capacity has been used in a number of protocols for genetic transformation via *A. tumefaciens*. However, the first protocols devised showed little efficiency for routine transformation. In addition, in many cases, transfer of the transgene to the progeny has not been demonstrated. A protocol giving a transformation efficiency of around 20% has been established. It is based on the co-cultivation of the bacteria with hypocotyl segments, development of calli and cell suspension cultures and finally regeneration of plants through induction of somatic embryogenesis. Using this method, not only plants harbouring the *NptII* and the *GUS* genes have been generated but also transgenic plants in which sucrose metabolism has been changed. Gilbert *et al.* using a similar protocol found that transformation efficiency was not influenced by explant age or binary plasmid, but was significantly influenced by the *Agrobacterium* strains, co-cultivation times, and cultivars. Nevertheless, three cultivars of carrots were transformed with a maximum efficiency of 12.1% with either acidic or basic chitinase.

### Sweet potato

Genetic transformation of sweet potato has been attempted using various gene transfer systems including electroporation of protoplasts, particle bombardment. *A rhizogenes* and *A. tumefaciens*. The generation of transgenic plants was reported only for *Agrobacterium*-mediated transformation. Previously established protocols for somatic embryogenesis have been used in some transformation procedures with *A. tumefaciens*. Newell *et al.* starting with root disks, obtained transgenic plants of the 'Jewel' cultivar expressing the GUS enzyme, the cowpea trypsin inhibitor and the snowdrop lectin with a transformation efficiency reaching 10%. Another protocol, also using embryogenesis but starting with apical meristems of the 'White star' cultivar, was described by Gama et al. with a similar efficiency. More recently Moran et al. established a protocol for the 'Jewel' cultivar using leaf disks and shoot organogenesis that gave 31.1 to 35.5% transformation efficiency, which is much higher than the protocol allowed the generation of plants expressing the *CryIIIA* gene of *B. thuringiensis*. The efficiency of this protocol needs to be tested on other cultivars.

### Allium species

Allium species are recalcitrant to transformation by *A. tumefaciens* although some of them, such as onion (*A. cepa* L.) is a host for *Agrobacterium*. In garlic (*A. sativum* L.) where gene transfer via breeding is limited by the sterility of commercial genotypes, biotechnology has great potential. In this species, an efficient method for callus culture and shoot regeneration has recently been established, but genetic transformation has not been published so far. The same situation prevails with leeks (*A. ampeloprasum* L.) for which embryogenic calli were induced in cell suspension cultures and regeneration competent protoplasts obtained.

## Resistance to Viruses

A programme devoted to the development of virus resistance in sweet potatoes has been initiated using the coat protein of the sweet potato potyvirus.

## Resistance to Insect and Fungi

The most important pests of sweet potatoes are insects, mainly the sweet potato weevil (*Cyclas formicarius*, Fab). Losses dues to insect attacks may reach 60 to 100%. The genetic of sweet potatoes is complex due to the hexaploid genome and self-incompatibility. Transfer of foreign genes via biotechnology is therefore of great interest. Moran et al. obtained several clones of the 'Jewel' sweet potato cultivar carrying the *CryIIIA* gene that exhibited some resistance to sweet potato weevil infestation under greenhouse and field conditions as compared to control plants although the level of expression of the gene was low. Extension to other cultivar and further field experiments are needed before commercial application. The transfer of two genes capable of conferring general resistance to insects, that encode the cowpea trypsin inhibitor and a mannose-specific lectin from snowdrop has been achieved both separately and in tandem.

Gilbert et al. generated transgenic plants of three carrot cultivars (Golden State, Danvers Half Long and Nanco) with either the acidic (petunia) or basic chitinase (tobacco). Two of the cultivars (Golden State and Nanco) were evaluated for the response to inoculation of detached petioles with a number of fungi. The rate of lesions was significantly lower in transgenic plants transformed with the acidic chitinase gene for *Botrytis cinerea*, *Rhizoctonia solani*, *Sclerotinium rolfsii* but remained unchanged for *Thielaviopsis basicola* and *Alternaria radicina*. These data are consistent with observations made in other plants showing that the efficacy of chitinase genes for enhancing disease tolerance in plants is variable according to the plant species, the type of chitinase and the type of pathogen.

## Quality Traits

The sugar content of carrots represents an important quality trait. Studies devoted to the elucidation of the role of enzymes involved in sugar metabolism and provision of a motive force for solute transport have been carried out. Transgenic carrots in which sucrose synthase, vacuolar and cell wall invertase and tonoplast $H^+$ ATPase have been repressed were generated. The transgenic plants had an altered phenotype with smaller roots and the strategy for improving the quality of the carrot (e.g. in terms of sugar content) remains to be established.

### Resistance to Abiotic Stresses

The generation of plants resistant to environmental stress is of great practical interest. Modifications of heat tolerance in carrots have been achieved by constitutively expressing or down-regulating a small heat shock protein gene, *Hsp17.7*. Constitutive expression resulted in an increase of heat tolerance while down-regulation resulted in lower tolerance. Practical evaluations of this strategy remain to be made.

### Root Vegetables as a Source of Genes and/or Promoters of Interest for other Crops

Sweet potato was used as the source of a trypsin inhibitor gene that can be used for generating insect resistance in cauliflower.

## Biotechnology of Leafy Vegetables (Cabbage, Broccoli, Cauliflower, Lettuce, Spinach) and Asparagus

This section is devoted to vegetables that are eaten as fresh or cooked leaves and to asparagus.

The *Brassica oleracea* (L.) species belong to the Brassicaceae family and include several important crops such as broccoli (*B. oleracea*, var *italica*), cauliflower (*B. oleracea*, var *botrytis*), cabbage (*B. oleracea*, var *capitata*), kale (*B. oleracea*, var *acephala*), Chinese kale (*B. oleracea*, var *alboglabra*) and Brussels sprouts (*B. oleracea*, var *gemmifera*). The Chinese cabbage belongs to another species (*B. campestris* ssp. *Pekinensis*).

Lettuce (*Lactuca sativa* L.) a member of the Compositae, and chicory (*Chicorium endivia* L), a member of the Asteraceae are high-value crops used in many countries as fresh leaf salads. In western Europe, Witloof chicories (*Chicorium intybus* L., var *foliosum* L.) are also used in salad as white buds, while root chicories (*Chicorium intybus* L. var *sativum*) are roasted and used as coffee surrogate.

Spinach (*Spinacia oleracea* L.) is a dioecious annual leafy vegetable of the Chenopodiaceae family.

Asparagus (*Asparagus officinalis* L.) is a monocotyledonous plant member of the Liliaceae cultivated as a herbaceous perennial and consumed as either white or green spears.

## Methods of Transformation

### *Brassica species*

Most research efforts have been directed at developing *A. tumefaciens*-mediated transformation, with little emphasis on direct transfer methods. A common procedure for all varieties of *B. oleracea* being unavailable, the protocols generally remain genotype-specific. *A rhizogenes*-mediated transformation causes formation of hairy roots that can be induced to form shoots. However, the bacteria formation of hairy roots that can be induced to form shoots. However, the bacteria carry the *rol* gene that may have phenotypic effects, for instance on flowering. The method has recently been proved efficient in transferring the *NptII* and other genes in 12 vegetable brassica cultivars representing six varieties: broccoli, Brussels sprouts, cabbage, cauliflower, rapid cycling cabbage, and Chinese cabbage. However, fertility was often reduced and morphogenic changes were noted in a number of plants. This method therefore needs further assessment and improvement before being used for commercial applications.

For cauliflower, a number of protocols were developed in the years 1988 to 1992 using marker genes. The protocol of De Block et al. has using a special combination of growth hormones, starting with cotyledons rather than hypocotyls, but still using silver nitrate, an inhibitor of ethylene action. Transgenic plants of three commercial genotypes were produced harbouring an antisense *Bcp1* gene encoding a protein essential for pollen functionality driven by a pollen-specific promoter. Another protocol, also using silver nitrate but starting with hypocotyls previously treated with 2,4-dichlorophenoxyacetic acid, gave very good regeneration rates and led to the production of over 100 primary transformants putatively harbouring a trypsin inhibitor gene conferring resistance to insects.

An *A. tumefaciens*-mediated transformations method has been devised for broccoli by Metz et al. using flowering stalks. This method, derived from the protocol of Toriyama et al., has been used for transferring *Bacillus thuringiensis* genes into broccoli with an efficiency of transformation of about 6.4%. An improved *A. rhizogenes*-mediated transformation of broccoli has been reported by Henzi et al. It gave 35% and 17% efficiency in the transformation of two cultivars, Shogun and Green beauty, respectively. This protocol has been used to generate plants producing low levels of ethylene, however, phenotypes altered due to the expression of the *rol* gene were often observed. This represents a serious limitation to the use of *A. rhizogenes*.

The Chinese cabbage (*B. campestris* ssp *pekinensis*) is considered as a recalcitrant species in plant regeneration. Procedures for the transformation and regeneration of transgenic plants have been reported by Jun et al. for the 'Spring Flavor' genotype and by Lim et al. for a number of other genotypes. However, the efficiency of the transformation was low and dependent upon the genotype.

### *Lettuce and chicory*

Michelmore et al. devised a routine protocol via *A. tumefaciens* enabling them to generate several hundred kanamycin-resistant plants, starting with cotyledon explants, that produced calli before regenerating shoots. Inheritance of the transgene was confirmed. This method has been recently used with success to generate tospovirus resistance in lettuce. A method for the generation of lettuce from adult leaf protoplasts and a protocol for transformation by electroporation has been published by Chupeau et al. that have not been retained in further studies. Rather, an *Agrobacterium*-mediated protocol starting with young leaves has been used for the introduction of a virus coat protein gene that lead to the generation of 16 primary transformants of three different cultivars showing accumulation of the protein out of a total of 87 putative transformants. Another protocol using *Agrobacterium* has been developed by Curtis *et al.*, starting from the cotyledons of seven-day-old seedlings. Overall, these data show that genotype-independent transformation procedures exist for the efficient transformation of lettuce. Transformation via *A. tumefaciens* of Witloof chicory has been achieved by Vermeulen et al. after optimization of shoot regeneration from leaf disks in order to confer herbicide resistance.

### *Spinach*

Although regeneration systems have been described from leaf disks, hypocotyl segments or root sections, very few reliable systems for transformation of spinach are available. Only two protocols have been recently described using *A. tumefaciens*-mediated gene transfer which have allowed stable transformation of spinach, one of them showing high efficiency. Also, expression of foreign DNA has been achieved in isolated spinach chloroplasts by electroporation, but this technical advance has no immediate biotechnological application.

### *Asparagus*

As mentioned by Conner and Abernethy, asparagus has been at the forefront of biotechnology developments in monocotyledonous plants, being the first such plant to be regenerated from tissue culture and isolated protoplasts and also to be genetically transformed. Transgenic asparagus plants have been generated by *A. tumefaciens*-mediated gene transfer through shoot regeneration from transformed calli or through embryogenesis. The efficiency of transformed calli or through embryogenesis. The efficiency of transformation was very low as in other monocotyledonous plants and difficult to use for practical applications. Direct DNA uptake by asparagus protoplasts has been achieved but the recovery of transgenic plants has not been reported. Due to the ease of regeneration of asparagus via embryogenesis, microprojectile bombardment may offer the most efficient approach for gene transfer. Li and Wolyn have generated transgenic plants expressing the *NptII* and *GUS* genes and a preliminary study of *GUS* transgene inheritance was performed. Cabrera-Ponce et al. also used the microprojectile bombardment method to transfer *hygromycin phosphotransferase*, *phosphinotrocin acetyl transferase* and *GUS* genes into embryogenic calli of asparagus. About 50 transgenic lines showing *GUS* expression were generated but inheritance studies are awaited.

It should be noticed that the integration of genetic engineering in an asparagus breeding programme is not easy. The most important cultivars are clonal hybrids that are genetically variable due to gene segregation among the progeny.

## Transformation for Herbicide Resistance

### *Cabbage*

Putative transformants of Chinese cabbage have been shown to express the *bar* (bialaphos resistance) gene. However, studies on the resistance of the transgenic plants to the herbicide and inheritability of the transgene have not been performed.

### *Lettuce*

Resistance to bialaphos has also been introduced into lettuce of the Evola cultivar by *A. tumefaciens*-mediated transformation.

Resistance to glufosinate was observed in axenic conditions and in the greenhouse and stable expression was confirmed over two generations. Field tests are awaited to further assess the advantages of the transgenic lines generated.

### *Chicory*

Transgenic Wiltloof chicories harbouring an *acetolactate synthase* gene have been generated and show resistance to the herbicide chlorsulfuron. Stable transformation has been observed in two selfed progeny, but large-scale tests in the greenhouse or the field are not reported.

### *Asparagus*

Five transgenic lines harbouring the *bar* gene and generated by particle bombardment were able to withstand the prescribed application of phosphinotrocin for weed control (0.5 to 1% solution by localized application). Large-scale and/or commercial applications are awaited.

## Resistance to Viruses

### *Cauliflower*

Transgenic cauliflower carrying the capsid gene and antisense gene VI of the cauliflower mosaic virus have been generated through *A. tumefaciens*-mediated transformation. However, while the transgenic of the transgenes was detected in all plants, the capsid protein was not present.

### *Chinese cabbage*

The tobacco mosaic virus 35S coat protein gene has been expected in five regenerants of the Spring Flavour cultivar of Chinese cabbage (*B. campestris*, ssp *pekinensis*). Stable inheritance of the gene was shown in the progeny, but virus resistance was not assessed.

### *Lettuce*

The tomato spotted wilt virus is a tospovirus transmitted mainly by the western flower thrips *Frankliniella occidentalis* to several hundred plant species, including the lettuce. Genetic sources of resistance often being limited, biotechnology has been considered as an alternative to conventional breeding. Transgenic lettuce plants, expressing the nucleocapsid protein gene of lettuce isolate of the virus, were protected against isolates of the virus not only when the protein accumulated at high levels, but also where transgene silencing occurred with high transcription rates and low steady-state mRNA levels. Confirmation of these protective effects under practical conditions and over several generations however, is still lacking.

Another virus, the lettuce mosaic potyvirus, can be destructive for lettuce crops. The coat protein of this virus has been introduced into three susceptible cultivars. The progeny of five transformants showed resistance to infection not only against the strain from which the coat protein originated, but also against other strains. However, the efficiency of resistance depended on the development stage of the plant at the time of inoculation. Although some plants (13%) showed stable resistance over the growth period, late viral infection was observed at advanced stages of development for most plants. Field tests need to be performed in order to evaluate the efficiency of protection in natural growing conditions and virus inoculation by aphids.

## Resistance to Insect and Fungi

### *Cauliflower*

Insect pests represent a serious problem for cauliflower cultivation. A trypsin inhibitor from the sweet potato has been transferred to Taiwan cauliflower cultivars that gave transgenic primary transformants substantial resistance to local insects upon *in planta* feeding bioassays. Progeny behaviour studies and field tests remain to be performed.

### *Broccoli*

Metz et al. have generated a large number of transgenic broccoli lines carrying the *Bt Cry1A(c)* gene, most of them causing 100% mortality of first instar larvae of the diamond moth, a major insect pest of crucifers. However, *Cry1A*-resistant larvae were able to survive on the transgenic plants. More recently, a synthetic *Bt Cry1C* gene was introduced also using the method developed by Metz. Lines producing high levels of *Cry1C* protein were protected not only from susceptible or *Cry1A* resistance to *Cry1C*. In addition, the *Cry1C*-transgenic broccoli were also resistant to other lepidopteran pests of crucifers such as cabbage looper and imported cabbageworm.

### *Cabbage*

The *Cry1A(c)* gene was likewise introduced into cabbage, however, the disadvantages of this gene in failing to control resistant insects is the same as already mentioned for broccoli. The introduction of other synthetic *Bt* genes is awaited in this variety.

### *Lettuce*

Transformation of lettuce with *A. tumefaciens* harbouring a maize *Ac* transposase and *Ds*, an empty transposon donor site has been used to generate mutants to lettuce that were screened for downy mildew resistance. This work represents a good example of the use of T-DNA mutagenesis combined with transposon tagging and genetic mapping with the aim of isolating genes of agronomic interest.

## Quality Traits

### *Brocccoli*

Transgenic lines of broccoli containing a tomato antisense 1-amino-cyclopropane-1-carboxylic acid oxidase gene, showed significant reduction of ethylene production in the florets. However, the fall in ethylene production was probably not enough, even though it sometimes reached more than 90%, to slow down senescence and preserve the quality of the florets. Higher levels of reduction are required to obtain interesting phenotypes, in particular through the use of homologous genes.

### *Lettuce*

Since lettuce accumulates high levels of nitrate and nitrate can be harmful to human health, great interest has been shown in reducing the nitrate content of the leaves. Besides tight control of cultivation conditions, the transfer of the nitrate reductase gene has been considered as an alternative approach. Curtis et al. have stably expressed a chimeric nitrate reductase gene of tobacco in transgenic lettuce. The level of nitrate was significantly reduced but not sufficiently to reach very low levels, especially in the older leaves. In addition, phenotypic alterations were observed such as chlorosis, dwarfing and early flowering. Further studies may render this strategy fully applicable at the commercial level.

Increasing the iron content of vegetables can have health benefits. An increase in the iron content of lettuce ranging from 1.2 to 1.7 times has been achieved by expressing a cDNA of soybean ferritine in lettuce via *A. tumefaciens* transformation. In addition, the transgenic lettuce had higher photosynthesis and growth rates, which represent interesting agronomic characters for commercial applications.

## Conclusion and Future Trends

Considering the relatively low economic importance of most of the crops dealt with here as compared to other major crops (soybean, potato, or tomato) and the fact that the first transgenic plants were generated less than two decades ago, it can be concluded that the research efforts reported are quite significant. In addition, the large number of field trials performed between 1994 and 1998 are a testimony to the efforts carried out mainly by private companies for the improvement of these so-called secondary or under-exploited species. It can be noticed, however, that field trials on cucurbits represent more than 60% of the total. Progress still remains to be made in a number of areas, including (i) the improvement of transformation protocols, (ii) the search and development of new genes of agronomical interest and (iii) the development of strategies to better meet with public acceptability of transgenic plants.

Most of these trends are common to all transgenic plants but some have higher relevance to the less important crops. In a number of cases, gene transfer methods have been developed that are restricted by variety or genotype. Also sometimes, the efficiency of the protocols is too low for practical applications that require the generation of a large number of transformation events. Efforts therefore remain to be made in the improvement of the transformation protocols, by increasing the efficiency of the regeneration, choosing the right strain of *A. tumefaciens*, defining the best conditions for direct transfer via particle bombardment and using an appropriate selectable maker gene. Also, the stability of transgene expression has not always been assessed. Proof of integrative transformation must be sought in genetic (transmission to the progeny, Southern blotting) and phenotypic (expression and effects of the transgene) data. The biotechnological approaches constitute an important supplement to conventional improvement programmes.

A combination of both genetic engineering and traditional breeding techniques is necessary for the genetic improvement of vegetable species. For example, some agronomically important traits such as virus tolerance have been dealt with by biotechnology and conventional breeding. Nevertheless, although dramatic progress has been made in genetic engineering, further efforts are needed to extend the number of species that will be engineered and the number of target genes to be used for protection against a wide range of diseases. This should lead to environmentally safer agricultural practices that will use fewer pesticides and will render genetic engineering better accepted by the consumer. Likewise, improving the nutritional and sensory quality is also a major objective. Targeting down-regulation or expression of genes at the right time and in the right tissue or organelle is one of the future challenges of biotechnology.

Although some tissue-specific promoters are already available, especially for legume seeds, there is a need to develop efficient promoters that will specifically drive gene expression in the part of the plant used for food (fruit, roots, leaves, etc.) or in specific organelles such as the chloroplast. The possible dispersion of antibiotic resistance genes, although theoretical, is of great concern to consumers. Methods are now available for removing selectable marker genes. However, so far they have mainly been applied to model plants. There is no doubt that, when extended to commercial products, they will contribute to overcoming public reluctance to accept genetically engineered food.

# Chapter 12

# *In vitro* Conservation of Fruits and Nuts

Many germplasm facilities for the preservation and distribution of fruit and nut germplasm are now instituting slow-growth and cryopreservation strategies. Some genera have well-defined methods, while the techniques for others are still under investigation. Primary collections of plant germplasm are often in field plantings that are vulnerable to disease, insect, and environmental stresses. Slow-growth techniques provide a secondary storage method for clonal field collections. Alternative germplasm storage technologies also provide storage modes for experimental material, allow for staging of commercial tissue culture crops, and provide a reserve of germplasm for plant distribution.

Cryopreservation in liquid nitrogen (LN) provides a low-input method for storing a base collection (long-term backup) of clonal materials. Recent improvements in cryopreservation methods make these long-term collections of clonal germplasm feasible. Both *in vitro* and cryopreserved collections provide insurance against the loss of valuable genetic resources and may provide alternative distribution methods. Medium-term storage of clonal plants involves slow-growth strategies such as temperature reduction, environmental manipulation, or chemical additions in the culture medium.

Storage techniques developed thus far provide several options so it is now possible to match improved techniques with a facility's needs and resources for the best possible plant preservation. New techniques and improvements in cryopreservation research have greatly increased the number of cryopreserved species. Suspension or callus cultures, dormant buds, *in vitro* grown apical meristems, isolated embryonic axes, seeds, somatic embryos, and pollen are now stored in LN. Cryopreserved collections of temperate plants of economic importance are now established in several countries. The storage of most temperate horticultural crops as base collections in liquid nitrogen is now feasible.

## Literature Review of Progress

### Medium-term Storage at Above Freezing Temperatures

*In vitro* collections play an important role in storing and distributing germplasm throughout the world. Certification programmes often incorporate *in vitro* culture as a standard technique for producing virus-negative plants from stock collections. *In vitro* culture systems are available for most temperate fruit and nut crops, but information on medium-term storage is limited for many genera. Most studies involve temperatures near freezing, but some tests of room temperature storage and chemical inhibition are also available. Published research is available for *Malus*, *Morus*, *Prunus*, *Punica* and *Pyrus*; however mot studies are restricted to a few genotypes and storage conditions. Published reports of *in vitro* storage systems for temperate nut trees are very limited; however, a species by species report of conservation methods currently applied to temperate tree fruit and nut crops is given below.

#### *Corylus*

More than 80 genotypes of hazelnut (*Corylus* sp.) *in vitro* cultures are stored at 4°C in low light (5 µmol $m^{-2}$ $s^{-1}$) at NCGR—Corvallis, Oregon, USA. Storage was also successful in total darkness where the mean storage duration for accessions held at 4°C in the dark is 1.26 years with a range of 8 months to 2.5 years.

#### *Malus*

Lundergan and Janick (1979) first suggested the feasibility of *in vitro* germplasm storage after successfully storing *Malus domestica* Borkh. cv. Golden Delicious *in vitro* for 12 months at 1°C and 4°C.m Success with other *Malus* species and cultivars included storage ranging from 9 months to 3.5 yr. Wilkins et al. (1988) studied cultures of five *Malus domestica* cultivars, *M. prunifolia* (Willd). Borkh., and *M. baccata* (L.) Borkh. and successfully stored them for 12 to 28 months at 4°C with a 16 hours photoperiod on an agar-based multiplication medium. Some rootstock cultivars were successfully stored on liquid medium. Charcoal in the medium increased storage time, but eliminated proliferation during storage. Two apple rootstock cultivars kept at 4°C in the dark stored better on medium with 6-benzylaminopurine (BAP) than on medium lacking the growth regulator. Storing plants immediately after subculture was important for the survival of multiple-shoot tufts, but shoot tips and nodal segments survived whether stored at 0, 10, or 20 days after subculture. Eckhard (1989) stored apple-shoot cultures at 2 to 4°C under low light for 1.5 years on a reduced sucrose, low BAP medium. Druart (1985) stored topped, partially submerged shoots of three *Malus* rootstocks and 'Golden Delicious' on hormone-free medium in the dark at 2°C for 1.5 to 3.5 years with 100 per cent survival. Apple species and cultivars (150 genotype) are stored at Changli Institute of Pomology (China) on modified MS medium at 2 to 4°C in the light (10 $\mu E\ m^{-2}\ s^{-1}$, 15 hour photoperiod) with yearly transfer.

***Morus***

*Morus nigra* L. shoot tips survived for only six months on multiplication medium at 4°C with a 16 hour photoperiod, but with activated charcoal in the medium, survival could be increased to 42 per cent after nine months at 25°C. Sharma and Thorpe (1990) stored 15 genotype of *Morus alba* L. for six months at 4°C in the dark on shoot proliferation medium (80 per cent viability).

***Prunus***

Marino et al. (1985) stored shoot cultures of three *Prunus* (peach, cherry) genotypes at 8°C, 4°C or –3°C for up to 10 months on multiplication medium. A 16 hour photoperiod was important for successful 4°C and 8°C storage for 90 days, but ten-month, dark storage at –3°C was better than under light for some genotypes. Cultures stored 14 days after subculture survived better than those stored immediately, and lower temperatures increased storage times. Druart (1985) stored 12 *Prunus* species and cultivars on basal medium at 2°C in the dark for up to four years; dimethyl sulphoxide or glycerol in the medium was toxic to the cultures. Survival of topped and partially submerged shoots was genotype dependent. Wilkins et al. (1988) stored five *Prunus* genotypes on multiplication medium at 4°C with a 16 hour photoperiod for nine to 18 months.

***Punica***

Pomegranate, *Punica granatum* L., shoot cultures died at 4°C, but survived for 18 months at 10°C with a 16 hour photoperiod.

***Pyrus***

*Pyrus communis* L. subsp. *caucasica* (Fed.) Browicz shoot tips grown on basal medium at 4°C, 8°C and 12°C with a 16 hour photoperiod exhibited depressed growth for 12 to 18 months, with highest survival at 4°C. Adding mannitol or increasing sucrose in the medium were not successful for storage at either 4°C or 28°C. *Pyrus pashia* D. Don had 100 per cent survival on multiplication medium at 4°C or 10°C with a 16 hour photoperiod after 12 months. *P. communis* cultivars La France, Bartlett, and La France × Bartlett had good survival at 5°C with a 16 hour photoperiod and at 1°C in darkness for 20 months, but 10°C and 15°C storage with light produced poor survival. Japanese pears *P. pyrifolia* (Burm.) Nakai. CVs. 'Shinsui', 'Nijisseiki', 'Shinchu', 'Kosui' 'Hosui' and 'Hakatado' were killed when stored at 5°C, 10°C, or 15°C with a photoperiod, but had 100 per cent survival when stored at 1°C in the dark for 12 months. ABA did not improve the survival of stored Japanese pears. Pears stored at NCGR-Corvallis in 1984 were kept in 20 × 100 mm tubes at 4°C in darkness, but tubes were replaced with tissue-culture bags in 1989. *Pyrus* accessions (169) were stored for eight months to 4.6 years with a mean storage time of 2.7 years in tissue-culture bags in the dark at 4°C. Storage of 46 genotypes with three treatments (4°C upright plants, 4°C three-quarters submerged, 1°C upright) showed genotype differences for the length of storage but few differences were noted among the three storage treatments.

**Long-term Storage in Liquid Nitrogen**

Cryopreservation of temperate fruit trees began in the 1970s when dormant bud freezing was successfully applied to apple, pear, peach, plum, and cherry; now additional techniques are available. Many temperate nut seeds are dehydration sensitive, liquid nitrogen sensitive, or survive for a year or less in 4°C storage. In this respect, they are similar to recalcitrant tropical seeds as demonstrated by Marzalina and Krishnapillay. Excised embryonic axes are excellent material for cryopreservation of wild populations, but cultivars require methods similar to those developed for fruit trees. The three major cryopreservation techniques—slow freezing, vitrification, and encapsulation—dehydration—are useful for these plant materials. Slow-freezing techniques developed in the 1970s by several investigators are used on many different species developed several plant vitrification solutions with highly concentrated cryoprotectants that allow cells to dehydrate quickly and cellular liquids to form glasses at low temperatures. Vitrification solution components, the duration of exposure, the size of plant material, the cryoprotectant toxicity, and the temperature of application are all important to plant survival. Dereuddre et al. (1990a, 1990b) devised a new cryopreservation system involving encapsulation of shoot tips in alginate beads followed by dehydration and direct exposure to LN. In addition, combinations of these techniques are also used in certain situations.

### *Carya*

Pence (1990) found most *Carya* embryonic axes dried to 5-10 per cent moisture content before being exposed to LN germinated or partially germinated, with some callus following thawing. Subsequent *in vitro* growth and development was best for fresh seed and declined from shoots to callus to no growth as the seed aged.

### *Castanea*

*Castanea* axes dried to about 8 per cent moisture before freezing produced callus upon recovery in initial testing. Chestnut embryonic axes desiccated to 20-30 per cent moisture before LN exposure had improved survival and some shoot formation.

### *Corylus*

Early studies by Pence (1990) found that cryopreserved embryonic axes and control axes of *Corylus* seeds produced only callus. Embryonic axes from fresh seed of *Corylus avellana* L. 'Morell' exhibited maximum recovery when axes were frozen at 12 per cent moisture content, while 'Butler' required 11 per cent moisture to obtain 50 per cent recovery. Whole seeds of *C. avellana* 'Barcelona' did not survive LN exposure following desiccation pretreatment, but embryonic axes were excised from the thawed seed and regrown in culture. Axes from stratified 'Barcelona' seed had improved shoot growth for both control and LN exposed treatments. Axes from stored, stratified seed dried to 8 per cent moisture were cryopreserved with 85 per cent viability and 70 per cent shoot growth, while only 30 per cent of unstratified axes produced shoots. Embryonic axes from seeds of *Corylus colurna* L., *C. americana* Marsh., and *C. sieboldiana* var. mandshurica (Maxim.) C. Schneider were stored in LN using this technique at NCGR—Corvallis and the National Seed Storage Laboratory, Ft Collins, Co. Regrowth of the thawed axes was 75-80 per cent for all three species.

### *Juglans*

Dried axes of *Juglans* seeds (5 per cent moisture) germinated or partially germinated producing shoots and/or roots *in vitro* following cryopreservation. Cryoprotectant with 5 M 1,2-propanediol and 20 per cent sucrose produced 75-91 per cent survival and regrowth of *Juglans* embryonic axes. Slow freezing *in vitro* grown shoot tips of walnut was also successful. Modified PVS2 cryoprotectant treatment combined with slow freezing (0.5°C/min) of shoot tips produced 34 per cent survival. Encapsulation-dehydration and slow freezing methods were successful with isolated walnut somatic embryos.

### *Malus*

Apple and pear winter buds exposed to subzero temperatures at slow freezing rates retained their viability after immersion in liquid nitrogen and apple buds taken from the shoots grew after being grafted onto rootstocks in the greenhouse (77 per cent regrowth). Dormant vegetative *Malus* buds cryopreserved using a combined dehydration-encapsulation technique had 80-100 per cent viability. In related studies winter-dormant buds had moisture contents ranging from 48 to 60 per cent. They required desiccation to 20-30 per cent moisture to survive LN exposure. Very few tolerant species, however, could be desiccated below 10 per cent. At maximum hardiness most buds were tolerant of desiccation. Genotypes that naturally tolerate desiccation and freezing to −30°C or colder at maximum hardiness would survive this procedure best. Cryopreserved dormant buds were either thawed slowly in room temperature air or rapidly in 40°C water. Pretreating dormant apple buds with sugars and other cryoprotectants enhanced the survival of less cold hardy taxa or those that do not sufficiently acclimate. Slow freezing below −10°C and immersion in LN without a cryoprotectant was successful with good regrowth *in vitro* after thawing for dormant-shoot tips from winter apple buds. Dormant buds of 500 apple genotypes are stored in the vapour phase of

LN at the National Seed Storage Laboratory (NSSL) in Ft Collins, Colorado. Single-bud sections from cold-hardened, dormant apple shoots are dried to 30 per cent moisture, cooled at 1°C/h to −30°C, held for 24 hours, and then stored in the LN vapour phase.

The first tests of cryopreservation on *in vitro* grown shoot tips of apple recovered only callus. Caswell et al. (1986) found that high sucrose concentrations in the culture medium improved the hardiness of *in vitro* grown apple shoots, but that the cold hardening response was genotype dependent. The effects of sucrose *in vitro* were applied to *Malus* cryopreservation by Niino and Sakai (1992) who used the encapsulation-dehydration method to cryopreserve *in vitro* grown apple shoot tips and obtained about 80 per cent regrowth after thawing. Decreases in moisture content of *in vitro* grown plants were also obtained through extended culture duration. Plants cultured for 70 days without transfer before cold acclimatization (CA) had more shoot formation following slow freezing and LN exposure than those cultured for 35 days. These results are attributed to lower meristem moisture contents and slowed shoot growth. A freezing rate of 0.1 to 0.2°C/min was suitable for *in vitro* grown apple shoot tips. Meristems of more than 70 cultivars are stored in LN at Changli Institute of Pomology, Hebei Academy of Agricultural and Forestry Sciences. Samples removed from LN after one month and one, two, and three years were re-cultured with no change in survival or plantlet regrowth.

Vitrification is also a successful technique for *Malus* shoot tips. Apple shoot tips dehydrated with PVS2 (30 per cent glycerol, 15 per cent ethylene glycol and 15 per cent DMSO in MS medium containing 0.4 M sucrose) at 25°C for 80 minutes produced 80 per cent shoot formation following vitrification. Zhao et al. (1995) studied the effects of plant vitrification solutions PVS1 to PVS5 on apple meristems; plant treated in PVS3 (50 per cent sucrose, 50 per cent glycerol) for 80 minutes before exposure to liquid nitrogen had the best regrowth.

***Morus***

Shoot tips of prefrozen winter buds of *Morus bombycis* Koidz. cv. Kenmochi survived immersion in liquid nitrogen, but grafts and cuttings did not survive. Wang et al. (1988) regenerated plants of *M. multicaulis* Loud. Cv. Lusang through shoot tip culture from frozen winter buds using a similar method. Niino et al. (1992b) demonstrated that excised shoot tips from winter buds of *M. bombycis* cv. Kenmochi, prefrozen to –20°C at 5°C/day produced more shoots than buds prefrozen at 10°C/day. Partially dehydrating the buds to about 38.5 per cent moisture content at 25°C prior to prefreezing to –20°C, improved the recovery rates. Alginate-coated, winter-hardened shoot tips of several *Morus* species had maximum shoot formation (81 per cent) when dehydrated to 22-25 per cent water content before freezing. Thirteen mulberry cultivars tested for cryopreservation as *in vitro* grown shoot tips produced survival ranging from 40 to 81.3 per cent with all methods tested: slow freezing (0.5°C/min to –42°C). vitrification (PVS2, 90 minutes), air-drying (24 per cent water content), and encapsulation-dehydration (33 per cent water content).

***Prunus***

Survival (70 to 80 per cent) of axillary apices excised from *Prunus persica* cv. GF 305 *in vitro* culture plants required pretreatment on a 5 per cent DMSO and 5 per cent proline culture medium before vitrification. This pretreatment medium was also effective and produced 69 per cent and 74 per cent shoot formation in two *prunus* rootstock cultivars. Dormant buds of ten *Prunus* species were cryopreserved with up to 100 per cent recovery by grafting. *In vitro*-grown *Prunus* shoots survived cryopreservation with 75 per cent regrowth when held at –30°C for 24 hours before being transferred to LN. Culturing on 14 per cent sucrose medium and chilling at 4°C enhanced the low-temperature tolerance of cryopreserved plantlets.

***Pyrus***

Dormant hardy *Pyrus* shoots were able to survive LN after prefreezing to –40°C or –50°C. Moriguchi et al. (1985) found that shoot tips form dormant buds of Japanese pear required prefreezing to –40 to –70°C before being exposed to LN. Oka et al. (1991) and Mi and Sanada (1992, 1994) recovered whole plants from cryopreserved buds.

*In vitro* grown pear-shoot meristems were first successfully cryopreserved in 1990. A slow-freezing method for *in vitro* grown pear meristems which incorporated cold acclimatization and slow cooling produced 55 to 95 per cent regrowth in cryopreserved shoot tips of four *Pyrus* species including a subtropical species, *P. koehnei*. Encapsulation-dehydration was applied to pear by

Dereuddre et al. (1990a, 1990b). A 0.75 M sucrose preculture and four hour dehydration (20 per cent residual water) produced 80 per cent recovery. A modified encapsulation-dehydration method developed by Niino and Sakai (1992) produced 70 per cent shoot formation for three pear cultivars. They applied the vitrification method to pears and obtained 40 to 72.5 per cent regrowth.

A comparison of slow freezing and vitrification methods using 28 *Pyrus* genotypes found that 61 per cent had better than 50 per cent regrowth following slow freezing (0.1°C/min), while only 43 per cent of the genotypes responded this well to the vitrification technique.

## Germplasm Storage

Fruit and nut trees in field genebanks are at risk from severe weather, insect and animal pests, and diseases. Quarantine laws designed to prevent the spread of diseases or insects also restrict global exchange of field germplasm. Although *in vitro* cultures are not necessary disease free, and may be virus infected or have bacterial contaminants, some countries allow *in vitro* cultures of satisfy quarantine restrictions. This makes *in vitro* collections valuable as complementary or secondary collections. Cryopreserved storage is the ultimate base storage; available for use in case of emergency but requiring little input of care or money.

### *In vitro* Stored Collections

Collections of temperate tree fruit and nut crops are held at various experiment stations and plant breeding centres throughout the world. *In vitro* stored collections are sometimes used as secondary collections but may also be the primary collection. Complete listings of *in vitro* stored germplasm collections are difficult to find, but germplasm workers in many countries use *in vitro* culture for other purposes including virus elimination.

### Cryopreserved Collections

A few countries have initiated clonal-germplasm storage is liquid nitrogen. Research is underway in many more countries. For temperate tree fruit and nut crops, cryopreserved collections are held in China (*Malus in vitro* grown meristems), Japan (*Cydonia*, *Malus* pollen, *Malus*, *Morus* dormant buds) and the USA (*Corylus* embryonic axes, *Malus* dormant buds, *Pyrus in vitro* grown meristems, *Corylus*, *Pyrus* pollen).

## Discussion: The Role of Storage Technologies

### *In vitro storage*

*In vitro* culture is an important tool for the international germplasm community, but much remains to be learned about optimal *in vitro* storage conditions. Many of the factors mentioned in individual research reports require more investigation. Important data are available on the size and type of propagule stored. Barlass and Skene (1983) found that single-rooted *Vitis* shoots respond differently from proliferating cultures. For apple, single-shoot tips, nodal segments, and cultures with multiple shoots respond differently to various storage conditions. Optimum age, size, and physiology must all be taken into account before cultures are stored.

Light quality and intensity are important for culture growth both before and during storage. In most cases experimentation into light effect is limited by lack of growth room availability in a facility. Data on culture storage in light versus darkness are available, but extensive information on the effects of light quality, duration, and intensity is not available for tree fruit and nut crops.

Culture conditions before storage, and culture time after subculture have important effects on storage time, but little has been done to study these conditions. The pre-storage culture period affected both storage length and the proliferation of *Prunus* rootstock culture following storage; 14 days was optimum. In some apple genotypes, proliferation was better following cold storage than in non-stored plants, but this also varied with genotype. Multiplication and storage media greatly affect the survival of cultures after storage. Storage time may be improved or limited by the growth regulators in storage media. For the most part this is a genotype dependent phenomenon. Research on the effects of growth regulators in storage media and the genetic stability of stored cultures is still limited. Genetic analysis of *in vitro* grown and stored plants is not a standard practice; genetic instability appears to be genotype dependent and it is difficult to generalize on its causes or probabilities. Field and molecular analyses are needed to determine genetic stability. Adventitious shoot production may be a cause of genotype variation during *in vitro* storage. Improvements in multiplication and storage media should reduce the likelihood of adventitious shoot production. Additional research is badly needed to develop standard techniques for genetic stability testing.

Contaminants are a major difficulty for any *in vitro* system. Slow-growing contaminants may persist without being noticed for long periods, then suddenly become evident during or after *in vitro* storage. Stored cultures may die from the debilitating effects of latent infections. Indexing cultures for latent bacterial and fungal infections should be a standard step in germplasm storage procedures. Bacteriological media used to detect cultivable contaminants are more effective than simply examining the cultures visually. Special methods are still needed to detect non-cultivable contaminants, such as obligate parasites. Healthier cultures, longer storage times, and safer materials for distribution can be assured through improved detection of bacterial contaminants. Germplasm storage in heat-sealed tissue culture bags can nearly eliminate fungal, bacterial, or insect contamination during the storage period.

### *Cryopreservation*

Cryopreservation techniques are now available for many forms of fruit and nut tree germplasm storage (cell suspensions, callus, shoot tips, somatic embryos, and embryonic axes). Some of these techniques are now used for long-term storage of germplasm. Future improvements in cryopreservation will require attention to several research topics. The choice of plant material is one important consideration since both growth stage and genotype affect survival following LN exposure. Response to cryopreservation techniques varies greatly with genotype; even related genotypes may have very different survival following LN exposure. The physiological status of mother plants directly impact survival following cryopreservation. More emphasis is needed on research into the physiological condition of plants prior to cryopreservation. Research into the physiology of cold acclimatization (CA) of cultures will be useful, especially since CA pretreatments are necessary for the success of many cryopreservation techniques.

Comparison of cryopreservation techniques remains difficult due to wide variations in the temperature, light conditions, and duration of CA used in different laboratories. Freezing rates and cryoprotectants have received much attention in the past but still require further study. Slow freezing, the first technique developed, remains an important protocol; a very slow freezing rate (0.1°C/ min to –40°C) produces the best survival for *in vitro* grown shoot tips from many species including *Pyrus* and *Malus*. Most *in vitro* systems require cryoprotectants, and highly concentrated solutions (such as PVS2) formerly used only with direct LN exposure are now used for some slow freezing and encapsulation methods.

Combined methods are successful for cryopreserving some difficult genotypes. In a combination of the slow-freezing technique and the encapsulation-dehydration technique, encapsulated grape axillary-shoot tips were slowly cooled before being exposed to LN, significantly increasing survival and shoot formation over encapsulation alone. Dehydration of encapsulated dormant apple buds with a vitrification solution, followed by LN exposure was also successful. Advances in cryopreservation of difficult genotypes may result from further exploration of combined techniques. No phenotypic changes have been observed in meristem derived plants of cryopreserved plant material. Genetic abnormalities due to cryopreservation are expected to be rare; however, more studies are needed to confirm the genetic stability of plants held in LN.

## Impact on the Storage and Distribution of Germplasm

### *In vitro* Storage

The use of *in vitro* stored plants as primary or secondary collections of clonal crops reduces the land area required for field genebanks. Three replicates per genotype are considered ideal for germplasm storage and using an *in vitro* culture for one or two of these replicates greatly decreases field space and labour coats. *In vitro* cultures are not guaranteed to be pathogen free; however virus-indexed materials can be stored *in vitro* to keep them in virus-free condition. Bacterial and fungal indexing can detect cultivable contaminants and provide propagules in which requesters can have a high degree of confidence.

*In vitro* cultures obtained from virus-elimination programmes and indexed for cultivable bacterial and fungal contaminants often meet phytosanitary requirements for import and export. Plants distributed as *in vitro* cultures are useful for many requesters and in many cases cultures survive international shipment better than traditionally propagated plants. A large percentage of the NCGR plant material is distributed as *in vitro* plantlets. Acclimatization of these plantlets requires the same care as other *in vitro* grown materials. Each plant shipment should include information on *in vitro* growth and acclimatization procedures.

### Cryopreservation

Cryopreservation is best suited for base (long-term) storage of clonal collections. The greatest cost of cryopreservation is in the initial storage of an accession, but very little input is needed for many years after storage. Collections in LN require little storage space: a 40 l dewar can hold as many as 3000 sample vials (i.e., five vials for each of 600 accessions). A clonal collection with representatives in the field, *in vitro*, and in LN would provide active, backup, and base storage for an accession for less cost and greater security over the long term than three field plants. Although plants can be distributed as cryopreserved samples, they are best kept as insurance in case of loss of actively growing accessions. Plants cryopreserved at one location can be shipped to a second location in a specially designed travel dewar. *Pyrus*, *Ribes* and *Rubus* meristems and *Corylus* embryonic axes cryopreserved in Oregon were shipped to Colorado by air freight in a travel dewar for base storage. For recovery the cryopreserved samples should be thawed by the recommended procedures, regrown *in vitro* into plantlets, and acclimatized to the greenhouse by techniques used for the specific plant type.

## Conclusions

Plant conservation and germplasm exchange using *in vitro* methods have increased over the past decade, mirroring perhaps the advances in research in this field. Improved global transportation and communication have led to a wider exchange of ideas as well as to the exchange of plant materials. More institutions are now taking advantage of improved techniques to provide *in vitro* base, primary, or secondary collections to protect their germplasm collections.

The advantages of *in vitro* conservation of important plant collections are the same as in the past, both in terms of phytosanitary considerations and plant security, but the willingness of curators to provide alternative storage for crops has increased. Further improvements to these techniques are of course always needed; research is needed to improve *in vitro* culture, storage and cryopreservation, including a myriad of aspects in each of these fields. Fortunately, *in vitro* culture and cryopreservation have progressed to the point where they can be used routinely in many laboratories. *In vitro* stored plantlets are used as primary or duplicate collections in several facilities. Cryopreserved samples for long-term (base) storage of important collections are now a reality as well. They provide an important, previously missing, form of germplasm storage for vegetatively propagated plants.

# Chapter 13

# NEW VARIETY OF FOOD CROPS

Throughout history, plant breeders have sought to be genetically modify food crops to improve yield and increase resistance to disease and plant pests. Initially these improvements were achieved by selecting seed from superior plants and reproducing these with continual selection and breeding. Traditional breeding methods have increased corn and wheat yields by approximately 100% over the last half century. However, traditional plant breeding methods are slow and unpredictable. To introduce a desired gene or set of genes by conventional breeding methods require a sexual cross between parental lines followed by repeated backcrossing between the hybrid offspring and one of the parents until progeny with the desired characteristics are obtained. Genes are only accessible from plants that can be sexually crossed, and many genes besides the desired gene(s) will be transferred. Biotechnology provides an opportunity to overcome some of the limitations of traditional breeding by enabling plant geneticists to identify and clone specific genes encoding desirable traits, such as protection against insect pests, and to introduce these genes selectively into already useful varieties of plants.

Sexual compatibility is no longer a limiting factor for transfer of desired traits. The transformation process is faster and more efficient because successfully transformed plants can readily be identified. Numerous traits are being assessed for their potential to yield products with the ability to (1) protect plants against various insect, fungal, and viral pests and plant pathogens; (2) provide selectivity to preferred herbicides; (3) improve agronomic performance such as crop yields; (4) increase nutritional value of food for humans and farm animals; (5) reduce naturally occurring toxicants, antinutrients, or allergens; (6) modify the ripening process to improve the flavor of fruits an vegetables; (7) use plants as factories to make environmentally friendly biodegradable polymers for packaging materials; and (8) use plants to produce pharmaceutical products more cost effectively, and so on. Since the initial reports of the first genetically modified plants 15 years ago, nearly all agronomically important crops have been genetically modified.

By the end of 1995, more than 70 different crop species had been transformed. At least 56 different crops have been planted in at least 34 countries and grown in more than 15,000 individual field sites. According to the U.S. Animal and plant Health inspection Service (APHIS), the most frequently field-tested traits are insect protection, virus protection, fungal resistance, herbicide tolerance, and food quality enhancements. The number of transformed crops and field trials will continue to expand as more new crop varieties are developed through biotechnology for eventual entry to the marketplace. At least 34 genetically modified plants have successfully completed regulatory review by the appropriate regulatory agencies. Genetically modified plants were grown commercially in 1996 on approximately 7 million acres in various world areas with approximately 30 million acres planted in 1997. Biotechnology provides plant breeders the opportunity to develop new varieties of food crops more efficiently and with greater potential benefit than has been possible with conventional breeding practices.

As with any technological innovation, there must be assurance that the technology will deliver food "as safe as" that developed through traditional breeding programs. This chapter addresses the safety assessment strategies for new varieties of food products developed through biotechnology. The approaches are consistent with the guidance developed by various international organizations such as OECD (1993, 1996, 1997), FAO/WHO (1996), and WHO (1991, 1995). An example of the application of these strategies to assess the safety of a genetically modified food plant. To understand how safety assessment approaches have evolved, it is instructive to review briefly how new traits are introduced into food crops through conventional breeding compared with biotechnology.

**Table 13.1. Examples of plant biotechnology products that have successfully completed regulatory review in at least one country**

| *Company* | *Genetic Trait* |
|---|---|
| AgrEvo Canada, Inc. | Glufosinate-tolerant canola |
| | Glufosinate-tolerant corn |
| | Glufosinate-tolerant soybean |
| Agritope, Inc. | Modified fruit-ripening tomato |
| Asgrow Seed Co. | Virus-resistant squash I |
| | Virus-resistant squash II |
| Bejo-Baden | Male sterility/glufosinate-tolerant chicory |
| Calgene, Inc. | Flavr Savr™ tomato |
| | Bromoxynil-tolerant cotton |
| | Laurate canola |
| China | Virus-resistant tomato |
| Ciba Seeds | Insect-protected corn |
| Cornell U./U. of Hawaii | Virus-resistant papaya |
| DeKalb Genetics Corp. | Glufosinate-tolerant corn |
| | Insect-protected corn |
| DNA Plant Technology | Improved ripening tomato |
| DuPont | Sulfonylurea-tolerant cotton |
| | High-oleic-acid-soybean |
| Florigene | Carnations with increased vasa life |
| | Carnations with modified flower colour |
| Monsanto | Improved ripening tomato |
| | Insect-protected potato |
| | Insect-protected cotton |
| | Glyphosate-tolerant cotton |
| | Glyphosate-tolerant canola |
| | Insect-protected corn |
| | Glyphosate-tolerant corn |
| Mycogen | Insect-protected corn |
| Northrup King | Insect-protected corn |
| Plant Genetic Systems | Male sterile oilseed rape |
| | Male sterility/glufosinate-tolerant corn |
| University of Saskatchewan | Sulfonylurea-tolerant flax |
| Zeneca/Petoseed | Improved ripening tomato |

## Introduction of Genes into Food Crops: Conventional Breeding Compared With Biotechnology

### Conventional Breeding

Genetic modification of plants has been practiced for hundreds of years with considerable success by plant breeders. Plant breeding has become a very sophisticated branch of applied genetics. Breeders have developed elegant procedures for crossing plants to introduce and maintain desirable traits such as increased yield and resistance to disease. With conventional breeding, cultivars highly adapted to cultivation can be improved by crossing them with closely related highly adapted cultivars to combine the desired features of the parents. If the parents are highly adapted to cultivation, their progeny tend to be highly adapted also. However, if a desirable traits is not available in highly adapted cultivars, the breeder will cross-adapt cultivars with cultivars from different geographic areas, more primitive varieties, or wild species.

The greater the diversity in genetic material of the parents, the greater the chance that undesirable characteristics (genes) will be introduced into the progeny. The progeny may be an "off-type", which means it has less desirable agronomic properties, such as stunted growth or poor yield, than parent cultivars. Eliminating plants with undesirable traits while retaining plants with desired features may require many backcrosses carried out over several generations. The greater the difference in genetic content of the parents, the more difficult sexual crossing can be. In vitro procedures such as embryo culture and protoplast fusion have made it possible to cross parents that may not be sexually compatible.

### Biotechnology

Biotechnology methods do not replace conventional breeding practices but can facilitate the introduction of desirable traits into the plant genome more efficiently and with greater precision. Unlike traditional breeding in which thousands of genes maybe introduced into progeny form their parents, only one or a few genes are typically introduced using biotechnological techniques. The source of the desired trait the breeder wishes to introduce into new crop varieties is no longer limited to those from sexually compatible species. This greatly expands the opportunity to introduce new traits to improve crop varieties.

The techniques used to introduce new traits into food crops vary depending on the kind of plant being transformed. Seed plants have been divided into two subclasses; monocotyledonous plants (monocots), whose seeds have a single cotyledon (meaning "seed leaf"), and dicotyledonous plants (dicots) or those with two cotyledons. Dicots (broadleaf plants like tomato, cotton, and soybean) were the first seed plants to have genes introduced via biotechnology. The first genetically modified plants were produced in 1982 using *Agrobacterium tumefaciens*, a bacterium that can transfer a portion of its own DNA (T-DNA) into the genome of plants. The disease-causing sequences from the *Agrobacterium* T-DNA have been deleted followed by insertion of DNA that encodes for a desired trait such as insect protection. Regulatory signals are added to enable the gene to function optimally in the plant. Border sequences in the plasmid delineate the desired genes that will typically be inserted in the plant genome.

The T-DNA is incorporated into a plasmid derived from *Escherichia coli*, which is transferred to *A. tumefaciens* via a conjugation process. *A. tumefaciens* containing the engineered plasmid is incubated with selected tissue from the host plant, and the desired genes are stably inserted into plant chromosome. The inserted genes can then be transferred to new plant varieties using traditional breeding methods. Monocots, which are represented by agronomically important crops such as wheat, rice, and corn, were not initially amenable to transformation with *A. tumefaciens*. Recently, more aggressive strains of *A. tumefaciens* with broader host ranges have been developed that have been used to introduce genes into some cereal crops. In addition, techniques such as protoplast transformation and particle bombardment have been used to transfer DNA directly into cells where it is stably inserted into the plant genome. These techniques have been particularly valuable in monocot plants for which the *Agrobacterium* transformation method was not effective or efficient.

In particle bombardment, very small metal beads (1-μ diameter) are coated with DAN and shot form a "gun" into the target monocot cell. Some of the plant cells will incorporate the desired gene(s) into their genome. Whether *Agrobacterium* transformation or particle bombardment is used to introduce genes into plat cells, only a small percentage of eligible plant cells will be successfully transformed. To identify the transformed cells in culture, genes for selectable markers have been included with the genetic information inserted into plant cells. The marker genes provide resistance to antibiotics, herbicides, and other substances added to the cell culture to inhibit the growth of nontransformed cells.

Plant cells that survive have been successfully transformed and can therefore be identified and regenerated into whole plants. As with conventional breeding, off-types are typically discarded. Progenies with normal agronomic properties that express the desired phenotype or trait are backcrossed with commercial crop varieties to generate seed bearing the new trait. The chances of success in identifying a progeny with the desired trait are much higher with recombinant DNA techniques because the breeder knows that the trait was successfully inserted into the plant genome. With conventional breeding, many generations of backcrossing may be required to identify a progeny with the desired trait.

## History of Safety Assessment for New Crop Varieties Developed by Conventional Breeding

During this century, our food supply has steadily improved in quality, variety, nutritional values, safety, and economy through the use of conventional breeding techniques to improve food crops. The history of safe use of new varieties of food crops developed by classical breeding techniques is based on several factors: (1) confidence and experience with the procedures used to generate new crop varieties; (2) knowledge of the composition of the food crop, including important nutrients and toxicants, if present; and (3) observation of the agronomic properties of new crop varieties that have been developed during this century via traditional breeding, only a very limited number of new varieties have presented safety concerns. The more well-known examples are (i) increased psoralens in certain varieties of celery that caused photo-dermatitis in food handlers; (ii) increased glycoalkaloid content in the Lenape variety of potatoes (glycoalkaloids can cause gastrointestinal discomfort); and (iii) increased cucurbatin levels in vegetable squashes that leave a bitter taste. In these three examples, the food crops contained endogenous toxicants affording protection against plant pests. The levels of these endogenous toxicants were inadvertently increased in these new varieties. These examples have caused plant breeders to monitor new varieties of food crops that naturally contain potentially harmful toxicants or antinutrients more carefully to be certain that levels of these substances are within acceptable limits. For example, the United States and Canada have set acceptable limits for glycoalkaloid levels in potatoes that new potato varieties must meet.

## Regulatory Oversight of New Crop Varieties

### *Conventional Breeding*

In the United States, there are no premarket regulatory requirements governing the introduction of new crop varieties developed through conventional breeding (with the exception that the variety must not exceed standards set for the level of certain natural toxicants such as glycoalkaloids in potatoes). Food safety is assessed through postmarketing measures under the Food, Drug, and Cosmetic Act. This practice is based on the long history of safe introduction of new crop varieties developed through conventional breeding. As discussed earlier, plant breeders monitor the quality of new plant varieties before they are introduced into commerce. In Europe and for some crops in Canada, new varieties of food crops must be registered with the government. This is not done for safety reasons but more as an assurance to the farmer that the new variety will perform at least as well as commercial varieties in the field. In the United States, the market place determines the performance acceptability of new crop varieties.

### *Biotechnology*

Because biotechnology is relatively new, there is considerably more regulatory oversight for the introduction of new crop varieties developed through this approach. In the United States, the regulatory authority to ensure the safety of food and feed products derived from plant biotechnology resides within the Food And Drug Administration (FDA). The U.S. Department of Agriculture (USDA) has the authority to ensure that genetically modified plants will not become plant pests. The Environmental Protection Agency (EPA) has the authority to evaluate the safety of plants that have been genetically modified for protection against plant pests such as insects, fungi, bacteria, and viruses.

The EPA also regulates herbicides by establishing herbicide tolerances for plants genetically modified to be herbicide tolerant. The movement and release of genetically modified plants in regulated by USDA under the Federal Plant Pest Act and the Plant Quarantine Act. Permits or notifications must be filed with USDA to obtain approval for field testing of new varieties of genetically modified plants under development. A determination that the genetically modified plant is not a plant pest (e.g., that the plants do not pose a risk to the environment or production agriculture) must be obtained before market introduction. The FDA has the regulatory authority to ensure the safety and wholesomeness of food and feed products, including those derived from genetically modified plants.

The FDA has adopted a decision tree approach to ensure safety of products derived from new varieties of food and feed developed by both traditional and newer genetic modification methods, including biotechnology. Under the Food Drug and Cosmetic Act the FDA has the authority to take regulatory action against a new variety of food if the genetic modifications would render the food "ordinarily injurious" to human health. Therefore, the FDA uses the same postmarket food adulterations approach and regulations for these products as are used for food and feed products derived from traditionally bred plant varieties. It is recommended that developers of genetically modified food consult with the FDA before commercialization, and this has been done with all genetically modified food products that are in the marketplace. The FDA has completed consultations on at least 29 different genetically modified crop plants to data.

The Federal Insecticide, Fungicide and Rodenticide Act (FIFRA) provides the EPA with the authority to regulate plants with bioengineered pesticidal traits. The EPA has approved at least seven different genetically modified plants with introduced pesticidal traits. These products are also reviewed by the FDA and USDA. In regard to regulation of genetically modified plants in other world areas. Health Canada regulates food safety, and Agri-Food Canada regulates feed and environmental safety as well as registration of specific plant varieties for certain crops. In Japan, the Ministry of Health and Welfare regulates food safety, whereas the Ministry of Agriculture, Food, and Fisheries regulates feed and environmental safety. In the European Union, environmental assessments are conducted before placing a product on the market, as outlined in the 90/220 EEC regulations.

The recently authorized Novel Foods Regulation governs food safety in the European Union Kingdom, Denmark, and the Netherlands. A Novel Feed Regulation is under consideration for overseeing feed safety under the current 90/220 EEC Process. Many other countries either have, or are in the process of developing, regulations for plant biotechnology products. The number of genetically modified plants that have successfully completed regulatory review in various countries include 23 in Canada, 5 in European Union, 15 in Japan, 3 in Mexico, 2 in Argentina, 1 each in Australia and Brazil.

## Safety assessment Strategies for New Crop Varieties Developed through Biotechnology

The safety issues that have been raised for genetically engineered plant products are similar to those for new varieties of plants derived from conventional breeding. For example, progenies derived from conventional breeding as well as biotechnology may have progenies with altered agronomic properties when compared with the parents. Varieties with altered agronomic properties are usually readily identified in field trials. It has been suggested that biotechnology may inadvertently cause the production of new toxicants through insertional mutagenesis events that activate dormant biosynthetic pathways. This scenario seems very unlikely—particularly for those crops where there has been considerable experience with traditional breeding. It is more likely that an insertional event would result in the increase or decrease in the expression of already recognized toxicants, which has very infrequently been observed in conventional breeding, as discussed above.

In conventional breeding, the genes bearing the desired trait have often not been identified. The same applies to the protein expression products of these introduced genes. Introduction of genes through biotechnology procedures in much more precise, for the genes have been defined before their introduction. However, unlike conventional breeding, introduced genes can be obtained from almost any source, not just sexually compatible relatives of the food crop. Therefore, as part of the safety evaluation, regulatory agencies have required a molecular characterization of any gene introduced into food crops. In addition, the protein expression product of the cloned gene must be characterized as to its function, specificity of action, and safety. There are no particular concerns about the safety of the genetic material itself. Genetic material form living organisms is made from the same four nucleotide building blocks; the only difference between genes is the nucleotide composition. The human gut at any one time has been estimated to contain hundreds of milligrams of DNA from ingested food and mucosal cells sloughed into the gastrointestinal tract. These DNA molecules are efficiently degraded by nucleotidases during digestion. The contribution of genetic material from genes cloned into food is trivial compared with the other sources of DNA in the gut. It was concluded that the introduced genes (DNA) in food products pose no more health risk to consumers than the rest of the DNA ingested from food sources.

***Molecular characterization***

For new plant varieties developed through biotechnology, the source of the gene introduced into the plant must be identified. The transformation system used to insert the gene into the plant genome must be defined as well as the number of copies of inserted genes and the integrity and stability of the genetic insert. For genes coding for pesticidal proteins, the EPA requires the following information: (1) description of the vectors, (2) identity of organisms used for the cloning of the vectors, (3) description of the methodologies used to clone the vectors, (4) vector description (Size in kilobases), (5) restriction endonuclease sites, (6) location and function of all relevant gene segments, (7) the final delivery system, (8) description of gene segments transferred to the plant, (9) description of whether the inserted genes are expressed constitutively or inducibly, (10) localization and expression of the pesticidal substance in plant parts, and (11) estimation of the gene copies inserted, and so on.

***Substantial equivalence***

A general consensus has existed among regulatory agencies in major world areas regarding the use of "substantial equivalence" as an approach to assess the safety and acceptability genetically modified crops (WHO 1995; OECD 1993; FAO/WHO 1996). This concept has been adapted in part from procedures that plant breeders have followed to monitor the acceptability of new varieties of food crops developed through conventional breeding.

According to OECD, "the concept of substantial equivalence embodies the idea that existing organisms used as food or food sources can serve as a basis for comparison when assessing the safety of human consumption of a food or food component that has been modified or is new. If a new food or food component is found to be substantially equivalent to an existing food or food component, it can be treated in the same manner with respect to safety, keeping in mind the that establishment of substantial equivalence is not a safety or nutritional assessment in itself, but an approach to compare a potential new food with its conventional counterpart". The use of substantial equivalence is considered a practical approach to assess the safety of new varieties of food. Safety is defined as a reasonable certainty that no harm will result from intended uses under the anticipated conditions of consumption. The following are the three possible outcomes relative to assessing the substantial equivalence of a genetically modified food or food component to a conventional counterpart.

1. The genetically modified food or food ingredient is substantially equivalent to the conventional counterpart.
2. The genetically modified food or food ingredient is substantially equivalent to the conventional counterpart with the exception of certain well-defined differences; or
3. The genetically modified food or food ingredient is not substantially equivalent to the conventional counterpart.

Examples of new varieties of food crops that could be considered substantially equivalent to conventional counterparts are virus-resistant plants produced by introduction of viral coat protein (viral protein is already present in plant tissue of conventional infected plants). Alternatively, if the food is processed removing any added traits and the composition has not changed (e.g., processed canola oil), the oil would be considered substantially equivalent to its conventional counterpart.

The second outcome listed above will apply to most genetically modified food crops. For example, many genetically modified crops will be substantially equivalent to conventional counterparts with the exception of a new trait that imparts a desired characteristic such as pest resistance. When this is the case, further safety assessment should focus on the new trait or well-defined differences. For some genetically modified crops or food components derived therefrom, it may not be possible to demonstrate substantial equivalence to a conventional counterpart, either because differences are not sufficiently well defined or because there is no appropriate counterpart with which to make a comparison. The absence of substantial equivalence does not imply that the genetically modified crop is any less safe. The safety assessment should focus on the nature of the changes (FAO/WHO 1996).

**Key Parameters for Assessment of Substantial Equivalence**

An assessment of substantial equivalence should focus on a comparison of agronomic characteristics combined with compositional analysis that includes key nutrients as well as toxicants and antinutrients that have potential health significance.

1. Agronomic traits are a good starting point for evaluating substantial equivalence of genetically modified plants to their conventional counterpart. These traits will normally be examined as an integral part of the development program leading to a new food plant variety. Agronomic traits are those characteristics measured by plant breeders for a given crop. For example, in the case of potatoes these may include yield, tuber size and distribution, dry matter content, and disease resistance. Agronomic properties may vary depending on local environmental conditions. Therefore, the genetically modified variety should be grown in the same geographical regions in which it will be grown commercially.
2. Key nutrients are those components in a particular food product that may have a substantial health impact in the overall diet. These may be major constituents (fats, proteins, carbohydrates) or minor components (essential minerals, vitamins). Critical nutrients to be assessed may be determined, in part, by knowledge of the function and expression product of the inserted gene (e.g., if an inserted gene expresses an enzyme that is involved in amino acid biosynthesis, the amino acid profile should be determined). Introduction of an invertase into potatoes could influence the carbohydrate metabolism; therefore, starch should be investigated. Alternatively, introduction of a storage protein high in methionine into the soybean could affect the amino acid content; therefore, the amino acid profile should be measured. Examples of the analysis of key nutrients in different genetically modified plants have been published.

   Given differences among consumption patterns and practices in various cultures and societies, the key nutrients to be examined may differ in various countries. The critical nutrients to be addressed should be determined using consumption data for the target region. For example, in Denmark, potatoes provide an important source of vitamin C in the diet (35%), not because of their high vitamin C content (20 mg/100 gm) but because of high consumption of potatoes (140 gm/day). In the United States, potatoes are not as important a source of vitamin C owing to lower potato consumption and the availability of other dietary sources of this vitamin.

3. Critical toxicants and antinutrients are those compounds known to be inherently present in a crop variety whose potency could have an impact on health if their levels were increased significantly (e.g., solanine glycoalkaloids in potatoes, trypsin inhibitors in soybeans). Knowledge of the biologic function of the protein expression product of the inserted gene could influence the decision about which toxicants or antinutrients to examine.

The analysis of important nutrients and toxicants or antinutrients should use validated or standard methods where available (e.g., methods from the Association of Official Analytical Chemists [AOAC] or other recognized bodies). When the genetically modified line is compared with the parental line or conventional varieties, the varieties should be brown under similar environmental and agronomic conditions. If there are no statistically significant differences in measured parameters between the genetically modified line and the parent or conventional lines, then the genetically modified variety can be considered substantially equivalent to the parent or conventional variety. If statistically significant differences are observed in measured parameters, then a comparison can be made with values available in literature or other sources. If the parameter for the genetically modified line is within the normal range for conventional varieties, further evaluation is not warranted. If the parameter for the genetically modified line is outside the normal range, further evaluation may be needed.

A toxicological or nutritional evaluation may be needed to asses whether the difference between conventional and genetically modified lines for a given parameter is biologically meaningful and requires further investigation. In the future, genetically modified lines for sexually compatible crops will be crossed using traditional breeding techniques. If substantial equivalence has been demonstrated for (1) corn with a gene producing insect protection and (2) corn with a gene for herbicide resistance, then crossing (1) and (2) will produce a new variety that is likely to be substantially equivalent to the parents. If there are no expected interactions between the two traits, then no additional evaluation will be needed. Potential genetic interactions will need to be considered on a case-by-case basis. For example, if two independent modifications to the same metabolic pathway are combined by traditional breeding, further analysis of the products of the metabolic pathway may be warranted.

**Safety Assessment for the Introduced Gene Expression Product**

Many genetically modified food crops have been shown to be substantially equivalent to conventional crops with the exception to the introduced trait(s) that may impart one or more characteristic such as pest resistance, selectivity to preferred herbicides, modification of the ripening process, and so forth. For these examples, the safety assessment should focus on the introduced traits, the protein expression product of the cloned gene. The developer should do the following:

1. Define the biological function, specificity, and more of action of the protein. If the protein is an enzyme, assess the potential effects of the enzyme on metabolic pathways and levels of endogenous metabolites based on its mode of action and specificity.
2. Compare the amino acid sequence of the protein to known sequences in protein databases to determine if the protein has sequence homology to food proteins, toxins, or allergens.
3. Assess the inherent digestibility of the protein in vitro with simulated gastric and intestinal protease preparations.
4. Determine the level of expression of the protein in the food. This effort should focus on the raw agricultural product or a specific processed food component (e.g., oil), as appropriate.

**Criteria for Concluding the Introduced Protein is "As-Safe-As" Proteins Already Present in Foods**

The following criteria are important in assessing the safety of a protein introduced into food or food components:

1. The protein has a history of safe consumption in other food crops.
2. The protein is functionally and structurally related to proteins with a history of safe consumption in food.
3. The biological function and specificity or mode of action of the protein raises no safety concerns.
4. The amino acid sequence of the protein is not similar to known protein allergens.
5. The protein is not derived from a food source with a history of allergy.

6. The amino acid sequence of the protein is not similar to known protein toxins or antinutrients.
7. The protein is susceptible to degradation by digestive enzyme.

For certain insect control proteins such as *Bacillus thuringiensis* (B.t.) family of proteins, regulatory agencies such as the EPA require administration of the protein as a single oral high dose to mice. The scientific rationale for an acute test in mice is that known protein toxins generally manifest their toxicity via acute mechanisms. The B.t. proteins have a long history of safe use (EPA 1988); as new forms of B.t. proteins are introduced, the EPA has stated that an acute dosing test in mice will provide assurance that the new B.t. protein varieties are safe. Such studies have been conducted for the Cry3A protein expressed in potatoes, the Cry1Ab protein expressed in corn, and the Cry1Ac protein expressed in cotton and the tomato.

Enzymes that provide selectivity against herbicides or function as selective markers have also been tested in mice. Acute testing may be recommended if the protein is derived form plants, or microorganisms that have no history of consumption. Acute toxicity testing of proteins is not normally needed if the protein (1) is not present in the food (E.g., removed during processing, as in vegetable oils), (2) has a history of consumption in food, and (3) is closely related functionally and structurally to proteins with a history of consumption in food. If safety testing of proteins is undertaken, it may be possible to isolate and purify the protein from the plant. However, this may not be practical if the protein is expressed in small amounts in plant tissue.

Alternatively, the protein could be produced in an appropriate fermentation system generating a protein that is biochemically equivalent to the protein produced in the plant. Proteins that meet the safety criteria listed and are nontoxic if administered to mice are considered "as-safe-as" other proteins naturally present in foods. No further safety or nutritional assessment of the new variety of food crop is necessary.

### Criteria for Concluding that Further Safety Evaluation is Required for the Introduced Protein

Some introduced proteins may require further safety and nutritional evaluation if they meet any of the following criteria.

1. The protein is functionally and structurally related to proteins that are known toxins or antinutrients.

2. The protein is derived from a food with a history of allergy.
3. The protein is structurally similar to known protein allergens.
4. The protein is not degraded by digestive enzymes.
5. The biological function of the protein has not been characterized and there are no structurally related proteins identified in protein databases.

If the introduced proteins meet one or more of the criteria listed above, additional testing may be warranted on the basis of a case-by-case assessment. Toxicology or nutritional end points that may need to be developed are summarized in the next section.

**Safety Testing Recommendations**

***Testing for toxicity and antinutritional effects***

Certain antinutritional proteins such as lectins or protease inhibitors are naturally present in plants and provide protection against insect pests. There has been an interest in inserting these proteins in food crops to provide an alternative to insecticides for control of insect pests, although the safety implications of such transformations have raised concerns. Some plant lectins exert antinutrients effects when fed to animals by binding to the brush-border epithelium of gut cells and thus disrupting nutrient absorption. If a lectin were to be introduced into a food crop to enhance protection against insect pests, it should be fed to animals to assess whether it acts as an antinutrient. Assessing its potential for binding to brush-border epithelium would be part of the safety evaluation. If the introduced lectin had antinutrient properties, a "no-effect-level" for antinutrient effects would have to be determined if an adequate safety margin existed.

Human nutritional studies may also be appropriate. Because some lectins are known food allergens, assessment of potential allergenicity will also be necessary, which will be discussed in the next section. A similar testing scheme for protease inhibitors can be envisioned. Any general safety questions regarding the introduced protein could be addressed by animal feeding studies designed to assess toxicity and nutritional endpoints. For proteins containing common amino acids, no additional testing for mutagenic, carcinogenic, or teratogenic potential is indicated because there is no evidence that proteins are genotoxic or that feeding proteins such as food enzymes have ever directly produced carcinogenic or teratogenic effects in laboratory animals.

### *Testing for allergenicity*

Further testing of an introduced protein for potential allergy would be indicated if (1) the protein were derived from a food source with a history of allergy or (2) the amino acid sequence of the protein matched (at least eight contiguous identical amino acids) that of a known protein allergen. The immunogenic potential of the introduced protein should be tested in one of the various solid-phase immunoassays such as the radio allergosorbent test (RAST) or RAST inhibition assay or the enzyme-linked immunosorbent assay (ELISA). Solid-phase immunoassays use IgE fractions of sera from individuals confirmed allergic to the food from which the gene coding for the introduced protein was derived. Where in vitro tests are negative or equivocal, then in vivo skin prick tests could be carried out. If no positive response was detected in the prick test, then double-blind placebo controlled food challenges with patients known to be allergic to the food could be carried out under controlled clinical conditions.

The in vivo studies would only be warranted if the gene were derived from one of the following eight food groups that account for more than 90% of all food allergies: eggs, milk, fish, crustacea, peanuts, soybeans, wheat and tree nuts. The testing scheme to detect allergens summarized above works effectively as demonstrated in the case of the Brazil nut 2S storage protein. This protein was introduced into soybeans to increase the sulfur amino acid content and thereby improve the bean's nutritional value for use in animal feeds. Because there are a small number of individuals allergic to Brazil nuts, it was decided to test sera from these individuals to see if their sera contained IgE that would cross-react with the Brazil nut 2S storage protein. Sera from eight out of nine Brazil-nut-allergic individuals reacted with this storage protein. Development of this soybean line containing the introduced Brazil nut storage protein as terminated. If the developer had wished to proceed with commercialization of this genetically modified soybean, all food derived from this soybean would have had to be labeled as containing Brazil nut protein. For many genetically modified plants, the protein expression product will be derived from a gene source that has no history of inducing allergy. If the amino acid sequence of the introduced protein dos not show sequence homology to known allergens, the physicochemical properties of the protein should still be assessed to determine if it shares the profile for allergens. These properties include (1) stability to food processing conditions (high temperature and pH changes that denature food proteins), and (2) resistance to gastric acidity and digestive proteases.

The level of expression of the introduced protein in food should also be determined because allergenic proteins often comprise a significant proportion of the local protein in that food. Regarding digestibility of proteins, it is well established that protein allergens tend to be resistant to digestion and generally constitute a significant percentage of the total protein (1-18%) in the food. Allergens are often stable to heat processing. These properties increases the potential that sufficient quantities of the protein will survive both food processing and the hostile environment of the gut to elicit immunologic reactions in the intestinal mucosa. However, there are undoubtedly examples of proteins in food that represent a small percentage of total protein, are not digestible, and do not elicit food allergy. According to allergenicity experts, there is currently no validated in vivo or in vitro model to predict allergenicity potential of proteins. There is therefore a need to develop a predictive model for allergenicity.

## Safety Considerations for Marker Genes

As stated earlier, marker genes are added to the genetic material inserted into plants to help identify successfully transformed plant cells. The use of marker genes and their proteins expression products in plant biotechnology has raised the following safety concerns: (1) potential horizontal transfer of the marker gene from plant cells to gut microflora that might reduce the efficacy of therapeutic antibiotics, (2) potential toxicity of the protein expression products of the marker gene, or (3) compromised antibiotic efficacy due to expression of the antibiotic marker gene product in the food. After detailed scientific review, it was concluded that the potential for horizontal gene transfer from plants to gut microflora is vanishingly small (FAO/WHO 1996; WHO 1993). Introduced genes are stably incorporated into the genome of plants, and there are no known mechanisms for direct transfer of genes from plants to micro organisms. Moreover, the DNA released from the plant during digestion is rapidly degraded. This degradation occurs well before the plant during digestion is rapidly degraded. This degradation occurs well before the plant material reaches the lower intestine, cecum, and colon, where gut microflora are found. The potential for the protein expression product of the antibiotic marker gene to compromise antibiotic efficacy is limited by (1) the digestibility of the expressed protein, (2) low expression of the protein in food, or (3) lack of available cofactors such as adenosine triphosphate (ATP) in the gastrointestinal tract that are required for antibiotic inactivation (FAO/WHO 1996). The safety assessment of the protein expression product of marker genes should focus on the approach outlined earlier for protein expression products of introduced genes (new traits).

**Assessment of Safety for Nonsubstantially Equivalent Genetically Modified Plants**

To date, there have been few, if any, examples of genetically modified plants that are not considered substantially equivalent to conventional counterparts (With the exception of well-defined differences in some cases). As discussed earlier, with future developments in biotechnology, products may be developed that have no conventional counterpart for which substantial equivalence can be assessed. These products could arise by transfer of genomic regions that have only partially been characterized (FAO/WHO 1996). Genomics is the study of all the genes of an organism and their organization into chromosomes, and it includes researching the links between gene structure an function, or expression, that are the foundations for bioengineering new plant products.

Desirable agronomic traits may result from the linkage of several genes functioning together to produce effects such as resistance to drought or alkaline soil conditions. Such a development would allow plants to survive and grow under harsh environmental conditions, and millions of acres of land that cannot be currently used to grow food crops would be available for cultivation of genetically modified food crops. In the future, there may be novel foods derived via biotechnology that have intended benefits, but further safety and nutritional evaluation will be needed before these foods can be marketed. In other cases, unintended changes in a new food crop may occur that could require further evaluation to understand what was changed and its significance to the safety and nutritional value of the food crop.

Toxicological and nutritional studies may be indicated in some cases to resolve safety and nutritional questions of food crops that are not substantially equivalent to their conventional counterparts.

If animal feeding studies are considered necessary, their objectives must be clear and the experimental design carefully planned to avoid nutritional problems that may confound data interpretation. There may be a need to modify existing protocols for testing whole food or food components—in particular by providing adequate nutrition with respect to diet. Nutritional imbalance may mask toxic effects. Incorporation of high levels of a food into the diet may also result in nutritional deficiencies that could cause adverse effects unrelated to the food being tested.

The usual concept of safety margins may not be applicable because is of often not possible to food a whole food at a high enough level to obtain anything close to a hundred fold margin of safety. The testing of specific components or extracts of a novel food may present on option for addressing safety and nutritional testing of the whole food. For certain foods, animals are not good models for predicting safety to man because animals exhibit species-specific sensitivity to the food. For example, dogs experience transient paralysis from consumption of macadamia nuts or develop fatal cardiac arrhythmias owing to sensitivity to theobromine in chocolate; male rats of one particular strain develop cardiomyopathy when fed diets high in vegetables oils.

Feeding animals high levels of various foods in the diet has produced anemia (rats and dogs fed onions), enlarged cecums (rats fed potatoes), mucosal lesions in the stomach (rats fed tomatoes or chilli powder), pulmonary emphysema (rats fed beans), and reduced breeding performance and survival (rats fed wheat flour). If nutritional quality is the important issue, it may be appropriate to go directly to nutritional studies in human volunteers rather than to use animal models that may have limited relevance to man. The industry currently uses "sip and spit" tests to evaluate the organoleptic properties of new varieties of vegetable crops as part of the quality assessment. Once the initial safety assessment has been completed for a nonsubstantially equivalent food product, sensory or nutritional evaluation or both, may be recommended to ensure the quality and acceptability of the product for consumers.

## Conclusion

Biotechnology provides plant breeders the opportunity to develop new varieties of food crops more efficiently and with greater potential benefit than has been possible with conventional breeding practices. To provide assurance that this technology will generate food "as safe as" that produced by traditional breeding programs, safety assessment strategies have been developed for products of plant biotechnology that have generally been accepted by international regulatory groups (FAO/WHO 1996). This strategy includes comparison of the agronomic properties and important nutrient-toxicant composition of genetically modified crops with their conventional counterparts. If the comparisons show no meaningful differences in the aforementioned parameters, the genetically modified food crop is said to be "substantially equivalent" to its conventional counterpart, and the safety and nutritional assessment is completed.

Many genetically modified foods will be substantially equivalent to conventional varieties with the exception of one or more introduced traits. The safety assessment will then focus on the introduced trait to determine if the protein expression product meets the criteria for being "as-safe-as" proteins already in the food supply. If protein expression products are derived from plants with a history of allergenicity; have a structural or functional similarity to known allergens, toxins, or antinutrients; or have functions that may alter the nutritional quality of the plant, then additional safety and nutritional studies may be needed. In the absence of validated assay to predict potential allergenicity, comparison of amino acid sequence homology of the protein with known allergens and assessment of its physicochemical properties are the best tools currently available to predict potential allergenicity. As biotechnology develops, future products may be developed that have no conventional counterpart for which substantial equivalence can be assessed. These varieties could be designed to have improved agronomic properties or provide important health and nutritional benefits.

The safety and nutritional assessment of these crops should be carried out on a case-by-case basis using the strategies outlined above. During the last few years, over 34 different genetically modified plant products have successfully completed regulatory review in countries around the world. The experiences gained in bringing these products to market have enabled developers and regulatory agencies to define the key safety questions and the science to answer the question. This progress will help provide guidance to the development of improved and safe plant biotechnology products for the future.

# Chapter 14

# SAFETY ASSESSMENT

The success of the genetic modification of plant species for insect and virus resistance and herbicide tolerance has led to the rapid development of improved commercial varieties for many crops. There are over 40 different genetically modified plant products that have been approved in at least one country, and about 30 million acres of genetically modified plant varieties were planted in 1997. One of these products the insect-protected corn even event MON 810 expresses the *Bacillus thuringiensis* proteins Cry1Ab. Yield Gard™ corn is protected from feeding damage normally caused by the European corn borer (ECB, *Ostrinia nubilalis*), southwestern corn borer (SWCB, *Diatraea grandiosella*), and pink borer (*Sesamia cretica*). A critical step in the commercial development of these products is gaining regulatory approvals for both production and import of the genetically modified plant products. Food, feed, and environmental approvals, and, in some countries, variety registration for new plant varieties, are necessary for genetically modified plants. This chapter is an overview of the information developed and provided to the regulatory agencies to obtain food, feed, and environmental approvals for Yield Gard™ corn.

Variety registration is the same for varieties developed through traditional breeding or biotechnology. The European corn borer (ECB, *Ostrinia nubilalis*) (Hubner) is an economically important pest of corn that has spread throughout the major corn-growing regions of the United States and Europe. Yield losses due to ECB damage are estimated to be 3 to 7% per borer per plant, which causes annual losses from \$37 to \$172/ha or corn. This insect typically has one to three generations per year. Insect damage to the plant includes lead feeding by the first generation; stalk tunneling by the first and second generations; and leaf sheath, collar feeding, and ear damage by the second and third generations.

Chemical control is ineffective once the borers have tunneled into the plant because the insect pest is then inaccessible to the chemical. The benefit of planting insect-protected corn include: (1) a reliable means to control these corn pests: (2) control of target insets while maintaining beneficial species; (3) reduced use of chemical insecticides; (4) reduced applicator exposure to chemical pesticides; (5) fit with integrated pest management (IPM) and sustainable agricultural systems; (6) reduced fumonisin (class of mycotoxins) levels in maize kernels no additional labor or machinery is required, allowing both large and small growers to maximize hybrid yield.

The average yield advantage measured during the first year of commercial planting of Yield Gard™ corn in the US was approximately 15%. The development of corn transformation methodology created the opportunity to protect corn plants from insect-feeding damage using genes isolated form the bacterium *Bacillus thuringiensis*. The cry1Ab gene was isolated form the *Bacillus thuringiensis kurstaki* (B.t.k.) HD-1 strain used in DiPel®, the leading microbial insecticide in agricultural use. Several laboratories have produced transgenic corn plants expressing proteins of the Cry1A class in corn.

In 1997, insect-protected corn varieties derived from the corn event MON 810, which express the cry1Ab protein, were commercialized in the United States. The food, feed, and environmental safety assessment data generated for corn line MON 810 has been used to obtain the regulatory approvals from the United States Environmental Protection Agency (EPA) and the United States Department of Agriculture (USDA) and to complete the consultation process with the Food and Drug Administration (FDA). Food, feed, and environmental approvals have also been obtained from the appropriate agencies in Canada and Japan. Approvals in Western Europe are nearing completion. Regulatory approvals continue to be a critical step in the expanding commercial launches of YieldGard™ corn in global markets.

The food, feed, and environmental safety assessment include product characterization consisting of molecular analysis of the inserted DNA, protein characterization, and protein expression levels; protein safety evaluation; compositional analysis of food components to establish substantial equivalence to commercial varieties and environmental assessment to ensure that there will be no deleterious effect on the environment.

## Molecular Characterization

The insect-protected corn line MON 810 was produced by microprojectile bombardment of embryogenic tissue with plasmid PV-ZMBK07. The PV-ZMBK-07 plasmid contain the *cry1Ab* gene, which produces the Cry1Ab protein that has insecticidal activity. The *Cry1Ab* gene from *Bacillus thuringiensis* subspecies HD-1 was modified to increase the levels of expression in plants. The enhanced cauliflower mosaic virus (CaMV) 35S promoter and hsp70 maize intron regulate the expression of the Cry1Ab protein. The 3' nontranslated region of the nopaline synthase (NOS) gene, isolated from the Ti plasmid of *Agrobacterium tumefaciens*, terminates transcription and directs polyadenylation of the mRNA.

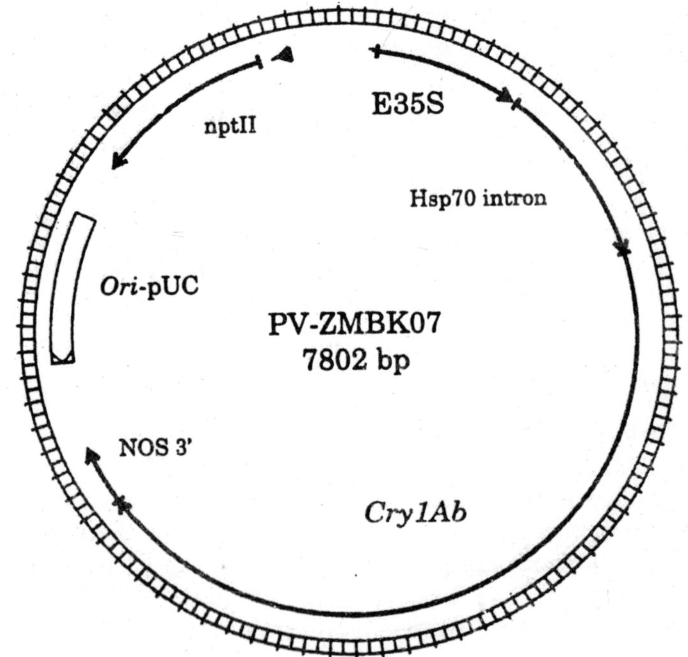


*Fig. 14.1. Plasmid used to produce insect-protected corn line MON 810.*

The plasmid contains the bacterial-selectable marker gene neomycin phosphotransferase (*nptII*), which confers resistance to amino-glycoside antibiotics (e.g., kanamycin). Southern blot analysis of corn line MON 810 demonstrated that a single copy of the *cry1Ab* gene was integrated into the corn genome. The *nptII* gene was not integrated during transformation. The *cry1Ab* gene inherited in the expected Mendelian pattern and is transmitted through pollen, which demonstrates stable integration into the nuclear genome. The integrity of the insert has been maintained during extensive breeding into commercial corn hybrids.

## Cry1Ab Protein Levels in the Corn Plant

The Cry1Ab protein expression levels were determined in order to: define the level of active ingredient for the EPA product label, calculate expected exposure levels, support the effective dose-insect-resistance management strategy, and demonstrate the stability of the product during breeding. The Cry1Ab protein expression levels were measured on samples form four different field trails: 1994 and 1995 trials in the United States and 1995 and 1996 trials in Europe. A direct double antibody sandwich enzyme-linked immunosorbent assay (ELISA) was developed and validated to quantify the levels of Cry1Ab protein in various plant tissues.

The Cry1Ab proteins levels in tissues collected from plants of line MON 810 have been consistent across several years of evaluation in the United States and Europe. The consistency of Cry1Ab expression through years of breeding supports the stability of the insert, which is an important component of product performance. The Cry1Ab protein levels in leaf samples collected during the season and whole plant samples are sufficient to provide effective protection from both first and second generation ECB feeding damage throughout the growing season.

## Safety Assessment of the Cry1Ab Protein

Safety assessment of the Cry1Ab protein includes protein characterization, digestion in stimulated gastric and intestinal fluids, acute oral toxicity evaluation in mice, and amino acid sequence comparison to known toxins and allergens. The use of *Escherichia coli*-produced Cry1Ab protein for safety assessment studies was justified by demonstrating the equivalence of the *E. coli*-produced and MON 810-produced protein.

**Cry1Ab Protein Mode of Action and Specificity**

The mode of action of *B. thuringiensis* delta-endotoxins such as the Cry1Ab protein has been studied extensively and reviewed. The Cry1Ab protein must be ingested by the susceptible insect to produce an insecticidal effect. Following ingestion, Cry1Ab toxins are solubilized and proteolytically processed to the active toxic core protein. After traversing the insect midgut peritrophic membrane, Cry1-type toxins selectively bind to specific receptors localized on the rush border midgut epithelium. Cation-specific pores are formed that disrupt midgut ion flow and thereby cause paralysis and death of the susceptible insect.

The Cry1Ab protein is insecticidal to only lepidopteran insects. Seven of the eighteen insects screened were sensitive to Cry1Ab and Cry1Ac proteins and all seven insects were, as expected, lepidopterans. This specificity is directly attributable to the presence of Cry1Ab specific receptors in the target insects.

There are no receptors for the protein delta-endotoxins of *B. thuringiensis* subspecies on the surface of mammalians intestinal cells; therefore, humans are not susceptible to these proteins. In addition to the lack of receptors for the *Bacillus thuringiensis kurstaki* (B.t.k) proteins, the absence of adverse effects in humans is further supported by numerous reviews on the safety of the B.t. proteins and a long history of safe use of microbial B.t. products.

**Digestion of Cry1Ab Protein in Simulated Gastric and Intestinal Fluids**

The trypsin-resistant core of the Cry1Ab protein was used in the simulated digestion study because this is the insecticidally active form of the Cry1Ab protein. In gastric fluid, the Cry1Ab protein degraded rapidly; more than 90% of the initially added Cry1Ab protein degraded within 30s of incubation in simulated gastric fluids, as assessed by Western blot analysis. The Cry1Ab protein bioactivity as measured by insect bioassay, also dissipated readily; 74 to 90% of the added Cry1Ab activity dissipated within 2 min during incubation in simulated gastric fluids, which was the earliest time point measured. To put the rapid degradation of the Cry1Ab protein in the stimulated gastric system into perspective, approximately 50% of solid food has been estimated to empty from the human stomach within 2 h, whereas liquid empties in approximately 25 min. In intestinal fluid, as expected, the Cry1Ab trypsin-resistant core protein did not degrade substantially after a 19.5-h incubation, as assessed by both Western blot analysis and insect bioassay. The tryptic core of this and other *B. thuringiensis* insecticidal proteins are widely known to be relatively resistant to digestion by serine protease like trypsin, a major protease in intestinal fluid.

**Acute Mouse Gavage Study with Cry1Ab Protein**

An acute mouse gavage study was performed to directly assess any potential toxicity associated with the Cry1Ab protein. An acute study was considered appropriate because toxic proteins are only known to exert acute effects. The Cry1Ab protein was administration by gavage to 3 groups of 10 male and female mice. The targeted doses of Cry1Ab protein administered to mice were 0, 400, 1000, and 4000 mg/kg. The highest dose embodied the maximum hazard dose concept of the Environmental Protection Agency. Bovine serum albumin (BSA) was gavaged at 4000 mg/kg to the control group. At the time of sacrifice, 7 days after doing, there were no statistically significant differences in mortality, body weights, cumulative body weight, or total food consumption between the BSA control groups and the Cry1Ab protein-treated groups. Results from this study demonstrate that the Cry1Ab protein is, as expected, not acutely toxic to mammals.

**Lack of Homology of Cry1Ab Protection to Known Protein Toxins**

One method for the assessment of potential toxic effects of proteins introduced into plants is to compare the amino acid sequence of the protein to known toxic proteins. Homologous proteins derived from a common ancestor are likely to share function. Therefore, it is undesirable to introduce DNA that encodes for proteins homologous to toxins. Homology is determined by comparing the degree of amino acid similarity between proteins using published criteria. The Cry1Ab protein does not show meaningful amino acid sequence similarity when compared with known protein toxins present in the PIR, EMBL, SwissProt, and GenBank protein databases with the exception of other B.t. proteins.

**Lack of Homology of Cry1Ab Protein to Known Allergens**

The most important factor to consider in assessing allergenic potential is whether the source of the gene being introduced into plants in allergenic. *B. thuringiensis*, the source of the *cry1Ab* gene, has no history of causing allergy. In over 30 years of commercial use, there have been no reports of allergenicity to *B. thuringiensis*, including occupational allergy associated with manufacture of products containing *B. thuringiensis*. In addition, the biochemical profile of the Cry1Ab protein provides a basis for allergenic assessment when compared with known protein allergens.

Proteins allergens must be stable to the peptic and tryptic digestion and the acid conditions of the digestive system if they are to pass through the intestinal mucosa to elicit an allergic response. Another significant factor contribution to the allergenicity of proteins is their high concentration in foods that elicit an allergic response. The physiochemical properties of the Cry1Ab protein are clearly distinct from these characteristics of known *allergens*. A comparison of the amino acid sequence of an introduced protein with the amino acid sequences of known allergens is also a useful indicator of allergenic potential, defined an immunologically relevant sequence comparison test for similarity between the amino acid sequence of the introduced protein and known allergens as a match of at least eight contiguous identical amino acids.

The amino acid sequences of the 219 allergens present in public domain genetic databases have been searched for similarity to the amino acid sequence of the Cry1Ab protein using the FASTA computer program. No allergen homologies or immunologically significant sequences were identified in the Cry1Ab protein. The two conclusions derived from this research were that (1) the *cry1Ab* gene introduced into corn does not encode a known allergen, and (2) the introduced protein does not share any immunologically significant amino acid sequence with known allergens. In summary, the Cry1Ab protein shows no amino acid similarly to known proteins toxins, other than B.t. proteins, and the Cry1Ab protein is rapidly degraded and its insecticidal activity lost under conditions that simulate mammalian digestion.

There were no indications of toxicity as measured by treatment-related adverse effects in mice to which Cry1A protein was administered by oral gavage. The *cry1Ab* gene was not derived from an allergenic source, does not possess immunologically relevant sequence similarity with known allergens, and does not possess the characteristics of known protein allergens. These studies support the safety of Cry1Ab protein, and are fully consistent with the extensive history of safe use for the Cry1Ab protein, which has high selectivity for insects with no deleterious effects on other types of organisms such as mammals, fish, birds, or invertebrates.

## Compositional Analysis of Corn Grain and Forage

The design of a food and feed safety assessment program for a genetically engineered crop requires detailed understanding of the uses of the crop and crop products in animal and human nutrition. Approximately 80% of the total quantity of corn grain is fed to livestock and poultry in the United States. Food and industrial uses account for the balance. Corn is palatable, readily digested by humans and by monogastric and ruminant animals, and is one the best sources of metabolizable energy among the grains. Corn is a major food and feed source worldwide; therefore, extensive data were generated to demonstrate the food and feed safety of the insect-protected corn line MON 810 New corn varieties developed by traditional breeding are typically selected for yield potential, and measurements of nutritional parameters are not routine.

Corn feed is supplemented with protein, minerals, and vitamins to meet the nutrient requirements of animals. Food safety can be demonstrated by confirming that the new food is substantially equivalent (i.e. as safe as) the conventional food. The establishment of substantially equivalence is an important component of a food safety assessment. Completed compositional analyses demonstrate that corn line MON 810 is substantially equivalent to current corn varieties. Compositional analyses were performed on grain

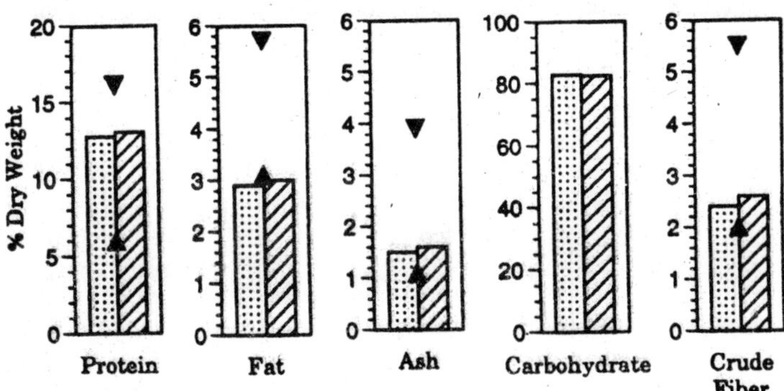


*Fig. 14.2. Proximate analysis of corn grain.*

harvested from the 1994 U.S. Field trials, forage and grain harvested from the 1995 European Union field trials. Grain and forage of corn line MON 810 were compared with those of the control line as well as published literature values.

The compositional parameters measured on grain samples form the 1994 U.S. field trials included proximate analyses (protein, fat, ash, crude fiber, and moisture), amino acid composition, fatty acid profile, calcium, and phosphorus. Carbohydrates were determined by calculations. Proximate analysis was performed on forage samples from the 1995 field trials in Europe. There were no statistically significant differences between the values for the control, MON 818, MON 810 for proximate analyses of the grain samples analyzed from the 1994 U.S. field trials. The values for the MON 818 and MON 810 lines were also comparable with the published literature. In the grain samples analyzed from the 1994 U.S. field trials, no statistically significant differences were observed in 10 of the 18 amino acids. Statistically significant difference were observed in 10 of the 18 amino acids.

Statistically significant increases were observed for eight amino acids (cystine, tryptophan, histidine, phenylalanine, alanine, proline, serine, and tyrosine). In the samples from the 1995 European

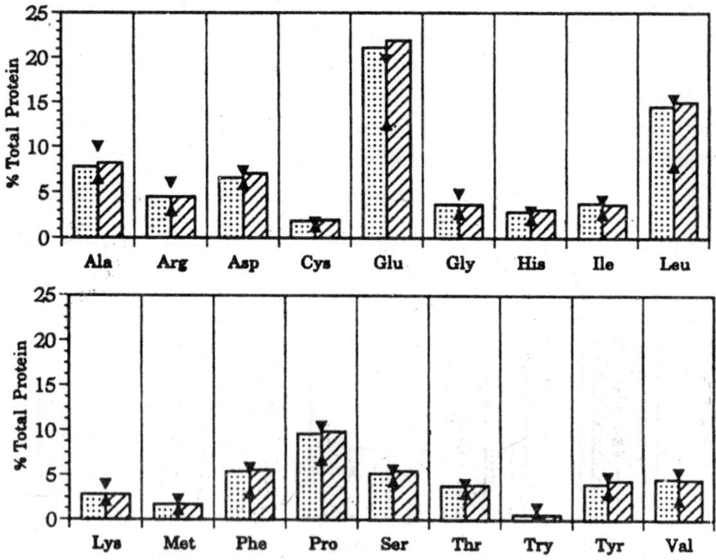


*Fig. 14.3. Amino acid composition of corn grain.*

field trial, no statistically significant differences for 16 of the 18 amino acids (data not shown) were observed. Methionine and tryptophan levels were lower in MON 810 grain from Europe compared with the control grain. The values for cysteine, histidine, and glutamic acid were slightly higher than the published literature range but similar to the nonmodified control. The few differences measured are minor, not consistent across multiple-year data, and are within either the published literature ranges or the values measured for the control. Therefore, these differences are not considered biologically meaningful. The observed variance could be due to environmental conditions, differences in analytical methodology, or differences from the older hybrids included in the published literature.

Ten fatty acids for which the measured values were near or below the limit of detection of the assay (arachidic, arachidonic, behenic, caprylic, capric, eicosadienoic, eicosatrienoic, eicosenoic, heptadecenoic, lauric, myristic, myristoleic, palmitoleic, and pentadecanoic) were excluded from the figure. No statistically significant differences were observed for the five fatty acids in the grain samples analyzed from the 1994 U.S. field trials. There were no statistically significant differences for four of the five fatty acids in grain from the 1995 European trials. Palmitic acid was slightly higher in MON 810 grain compared with the control grain. The higher palmitic acid value was not consistent across multiple-year data within published ranges; therefore, it was not considered biologically meaningful.

Statistically, the calcium level in MON 810 and MON 818, but this small difference was not considered biologically meaningful.

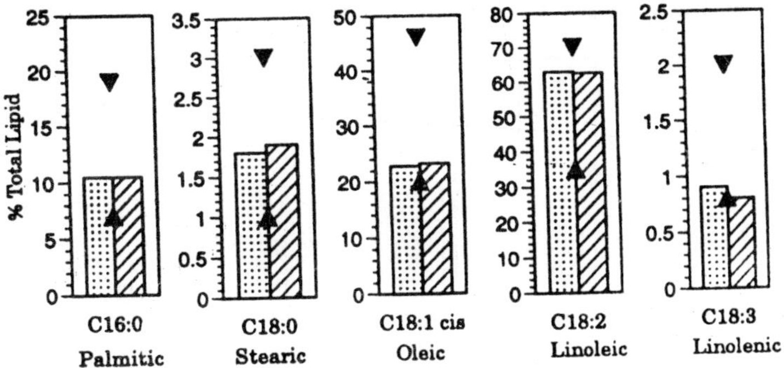


*Fig. 14.4. Fatty acid analysis of corn grain.*

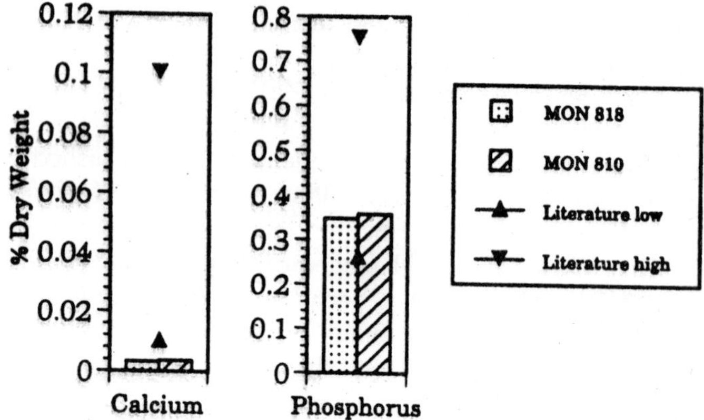


*Fig. 14.5. Calcium and phosphorus analysis of corn grain.*

The calcium values for both MON 810 and MON 818 are below the published literature range. The observed difference from literature values could be due to variances in hybrids included in the analyses, differences in analytical methodology, or both. No statistically significant differences in the phosphorus values between the MON 810 line and the control line MON 818 were noted.

There were not statistically significant differences in the values for fat, ash, neutral detergent fiber, acid detergent fiber, carbohydrate, and dry matter content between the MON 810 line and the control line MON 820. Statistically, the protein level was significantly increased in forage of MON 810 compared with the control protein value, but this was not considered biology important. The observed ranges are within the published literature ranges. The grain and forage compositional data confirmed that corn line MON 810 is substantially equivalent to the parental hybrid as well as traditional corn hybrids. Processing is unlikely to alter the

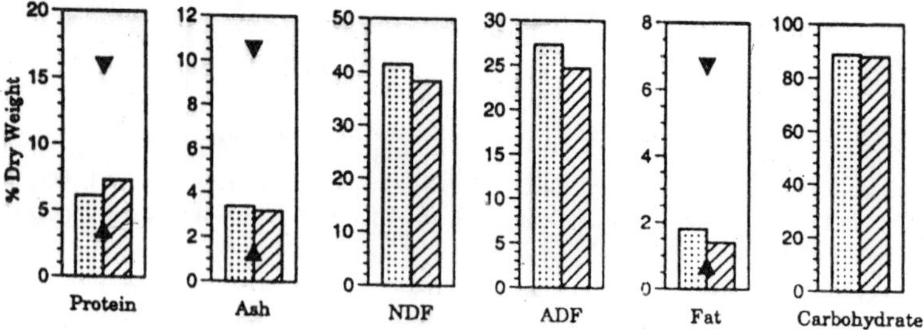


*Fig. 14.6. Analysis of forage from four field sites.*

compositional components of corn; therefore, products derived from corn grain will also be substantially equivalent and as safe as current corn-hybrid-derived products. Similar data have been published to confirm the substantial equivalence of several genetically modified crops: Roundup Ready™ soybean; New Leaf™, Bollgard™ cotton, Roundup Ready™ cotton; and Bt-tomatoes.

## Environmental Impact of the Product

There is extensive information on the lack of nontarget effects from microbial preparations of B.t.k. strains containing the Cry1Ab protein. The full-length Cry1Ab protein encoded by the *cry1Ab* gene used to produce the insect-protected corn plants and the insecticidally active core protein produced in the insect gut following ingestion are identical to the respective full-length and trypsin-resistant core Cry1Ab proteins contained in microbial formulations that have been used safely for over 30 years.

The B.t.k. Cry1A proteins are extremely selective for the lepidopteran insects, bind specifically to receptors on the midgut of lepidopteran insects, and have no deleterious effect on beneficial nontarget insects, including predators and parasitoids of lepidopteran insect pests or honeybee (*Apis mellifera*). To confirm the expand on the preceding results produced for the microbial products containing the same Cry1Ab protein as produced in YieldGard™ corn, the potential impact of the Cry1Ab protein on nontarget organisms was assessed on several representative organisms. These studies were conducted with the trypsin-resistant core of the Cry1Ab protein because this is the insecticidally active portion of the protein.

The nontarget insect species included larvae and adult honey been (*Apis mellifera* L.), a beneficial insect pollinator; green lacewing larvae (*Chrysopa carnea*), a beneficial predatory insect; hymenoptera (*Bachymeria intermedia*), a beneficial parasite of the housefly; the ladybird beetle (*Hippodamia convergens*), a beneficial predaceous insect; and earthworms (*Eisenia fetida*). Leaf material of MON 810 plants was used for the Collembola (*Folsomia candida*) nontarget soil organism study. Because of the potential exposure of aquatic invertebrates to corn pollen containing the Cry1Ab protein, a toxicity test was performed with *Daphnia magna*. The mortality of nonlepidopteran insect species and three other representative organisms exposed to the Cry1Ab protein did not differ significantly from control mortality. The no-observed effect level (NOEL) was the highest concentration tested.

In addition to these studies, no acute detrimental effects were observed for three predator species (*Colemegilla maculata*, *Orius insidious*, and *Chrysoperla carnea*) exposed to pollen of plants expressing the Cry1Ab protein. No effects were observed when *Folsomia candida* and *Oppia nitens* were fed cotton leaf material containing the Cry1Ac and Cry1Ab proteins. These results demonstrate the safety of the Cry1Ab protein to nontarget organisms. In addition to assessing the potential effect of the Cry1Ab protein on nontarget organisms, a study was conducted to confirm the expected rapid degradation of the Cry1Ab protein in soil. Corn plant tissues remaining after harvest may be tilled into the soil or remain on the soil surface (no till), depending upon agricultural practices following harvest.

The degradation rate of the Cry1Ab protein was assessed by measuring the decrease in insecticidal activity of transgenic corn tissue incubated in soil. The Cry1Ab protein, as a component of corn tissue, had an estimated $DT_{50}$ (time to 50% reduction of bioactivity) and $DT_{90}$ (time to 90% reduction of bioactivity) of 1.6 and 15 days, respectively. This measured rate of degradation in soil is comparable to that reported for the Cry1Ab and Cry1Ac proteins in genetically modified cotton and to the degradation rate reported for microbial B.t. products. This rapid degradation further supports the lack of deleterious effects on nontarget soil organisms. Entomologists have observed that insect populations adapt to insecticides if those insecticides are not managed correctly. Integrated pest management (IPM) was developed as result of industry experiences with chemical insecticides.

The components of the insect resistance management strategy implemented to maximize the sustainability of YieldGard™ corn include (1) expanding the knowledge of insect biology and ecology, (2) using an effective dose product that kills nearly all of the resistant heterozygote insects, (3) creating refuges to support populations of Cry1Ab-susceptible insects, (4) monitoring for any incidents of pesticide resistance and implementing a containment plan, (5) employing IPM practices that encourage ecosystem diversity and multiple tactics for insect control, (6) educating growers to ensure implementation of these strategies, and (7) developing products with alternative modes of action.

## CONCLUSION

The MON 810 plants have demonstrated effective control of the targeted insect pests in corn field trials since 1993. The *cry1Ab* gene has been crossed into commercial corn inbreds to produce hybrids of superior agronomic performance with resistance to lepidopteran insects. Detailed food, feed, and environmental safety assessments have confirmed the safety of this product and supported the regulatory approval of insect-protected corn line MON 810. The analyses included (1) detailed molecular characterization of the DNA introduced, (2) safety assessment of the expressed Cry1Ab protein, (3) compositional analysis of corn grain and forage, and (4) environmental impact assessment of the corn plants. These studies demonstrated that the Cry1Ab protein is safe to nontarget organisms, including humans, animals, and beneficial insects. Additionally, MON 810 plants and grain were shown to be substantially equivalent to (i.e. as safe as) conventional corn varieties. These data continue to be used to obtain regulatory approvals from the appropriate regulatory agencies around the world. This safety assessment approach has been used for over 40 different genetically modified plant products.

# INDEX

**A**



**B**